REVISE EDEXCEL GCSE
Mathematics
Specification A Linear

REVISION GUIDE

Higher

Series Director: Keith Pledger

Series Editor: Graham Cumming

Authors: Harry Smith, Gwenllian Burns, Jean Linsky

A note from the publisher

In order to ensure that this resource offers high-quality support for the associated Pearson qualification, it has been through a review process by the awarding body. This process confirms that this resource fully covers the teaching and learning content of the specification or part of a specification at which it is aimed. It also confirms that it demonstrates an appropriate balance between the development of subject skills, knowledge and understanding, in addition to preparation for assessment.

Endorsement does not cover any guidance on assessment activities or processes (e.g. practice questions or advice on how to answer assessment questions), included in the resource nor does it prescribe any particular approach to the teaching or delivery of a related course.

While the publishers have made every attempt to ensure that advice on the qualification and its assessment is accurate, the official specification and associated assessment guidance materials are the only authoritative source of information and should always be referred to for definitive guidance.

Pearson examiners have not contributed to any sections in this resource relevant to examination papers for which they have responsibility.

Examiners will not use endorsed resources as a source of material for any assessment set by Pearson.

Endorsement of a resource does not mean that the resource is required to achieve this Pearson qualification, nor does it mean that it is the only suitable material available to support the qualification, and any resource lists produced by the awarding body shall include this and other appropriate resources.

For the full range of Pearson revision titles across GCSE, BTEC and AS/A Level visit:
www.pearsonschools.co.uk/revise

ALWAYS LEARNING

PEARSON

Contents

A small bit of small print

A grade allocated to a question represents the highest grade covered by that question. Sub-parts of the question may cover lower grade material.

The grade range of a topic represents the usual grade range that the topic is assessed at. The topic may form part of a higher grade question if tested within the context of another topic.

Questions in this book are targeted at the grades indicated.

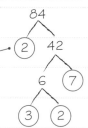

Factors and primes

The FACTORS of a number are any numbers that divide into it exactly.

A PRIME NUMBER has exactly two factors. It can only be divided by 1 and itself.

Prime factors

If a number is a factor of another number AND it is a prime number then it is called a PRIME FACTOR. You use a factor tree to find prime factors.

Remember to circle the prime factors as you go along. The order doesn't matter.

$$84 = 2 \times 2 \times 3 \times 7 \longrightarrow \text{Remember to put in the multiplication signs.}$$
$$= 2^2 \times 3 \times 7 \cdot \longrightarrow \text{This is called a PRODUCT of PRIME FACTORS.}$$

The highest common factor (HCF) of two numbers is the HIGHEST NUMBER that is a FACTOR of both numbers.

The lowest common multiple (LCM) of two numbers is the LOWEST NUMBER that is a MULTIPLE of both numbers.

Worked example

grade **C**

(a) Express 108 as a product of its prime factors.

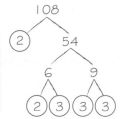

Everything in red is part of the answer.

$108 = 2 \times 2 \times 3 \times 3 \times 3 = 2^2 \times 3^3$

(b) Find the highest common factor (HCF) of 108 and 24

$108 = ② \times ② \times ③ \times 3 \times 3$
$24 = ② \times ② \times 2 \times ③$
HCF is $2 \times 2 \times 3 = 12$

(c) Find the lowest common multiple (LCM) of 108 and 24

$LCM = 12 \times 3 \times 3 \times 2 = 216$

When drawing a factor tree, make sure you continue until every branch ends with a prime number. It doesn't matter which factor pairs you choose for each branch. At the end, don't forget to write out the product of primes.

Check it!
$2 \times 2 \times 3 \times 3 \times 3 = 108$ ✓

To find the HCF circle all the prime numbers which are **common** to both products of prime factors. 2 appears twice in both products so you have to circle it twice. Multiply the circled numbers together to find the HCF.

To find the LCM multiply the HCF by the numbers in both products that were not circled in part (b).
Alternatively, you can multiply 108 and 24 together and divide by the HCF:
$108 \times 24 = 2592$
$2592 \div 12 = 216$

Now try this

edexcel

(a) Express the following numbers as products of their prime factors.
 (i) 60
 (ii) 96   **(4 marks)**

(b) Find the highest common factor (HCF) of 60 and 96   **(1 mark)**

(c) Work out the lowest common multiple (LCM) of 60 and 96   **(1 mark)**

grade **C**

A* A B C D

Indices 1

 Index laws

Indices include square roots, cube roots and powers.

You can use the index laws to simplify powers and roots.

$a^m \times a^n = a^{m+n}$

$4^3 \times 4^7 = 4^{3+7} = 4^{10}$

$\dfrac{a^m}{a^n} = a^{m-n}$

$12^8 \div 12^3 = 12^{8-3} = 12^5$

$(a^m)^n = a^{mn}$

$(7^3)^5 = 7^{3 \times 5} = 7^{15}$

② Cube root

The cube root of a positive number is positive.

$4 \times 4 \times 4 = 64$

$4^3 = 64$

$\sqrt[3]{64} = 4$

The cube root of a negative number is negative.

$-4 \times -4 \times -4 = -64$

$(-4)^3 = -64$

$\sqrt[3]{-64} = -4$

③ Powers of 0 and 1

Anything raised to the power 0 is equal to 1

$6^0 = 1 \quad 1^0 = 1 \quad 7223^0 = 1 \quad (-5)^0 = 1$

Anything raised to the power 1 is equal to itself.

$8^1 = 8 \quad 499^1 = 499 \quad (-3)^1 = -3$

Indices checklist

The base numbers have to be the same.

If there's no index, the number has the power 1

Be careful with negatives: $(-3)^2 = 9$

Worked example

grade **C**

Work out the value of

(a) $\dfrac{6^3 \times 6}{6^2}$

$\dfrac{6^3 \times 6^1}{6^2} = \dfrac{6^{3+1}}{6^2} = \dfrac{6^4}{6^2} = 6^2 = 36$

(b) $100^0 + 2 \times \sqrt[3]{-27}$

$100^0 + 2 \times \sqrt[3]{-27} = 1 + 2 \times (-3)$

$= 1 + (-6)$

$= -5$

Use the index laws to simplify your calculations as much as possible.

Always use the correct order of operations:

Brackets Indices Divide Multiply Add Subtract

Tip!

1. 'Work out the value of' means 'evaluate' so the final answer must be a number.

2. You need to be able to remember the cubes of 2, 3, 4, 5 and 10 and their corresponding cube roots.

Now try this

 grade **C**

1. (a) Work out the value of $3^6 \div 3^2$ **(1 mark)**

   (b) Work out the value of $\dfrac{(2^4)^5}{2^{17}}$ **(1 mark)**

grade **C**

2. (a) Simplify $(a^2)^4$ **(1 mark)**

   (b) Work out the value of x when
   $2^{30} \div 8^9 = 2^x$ **(2 marks)**

 edexcel

 grade **A**

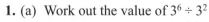 In part (b), convert 8 to a power of 2 and then use the index laws.

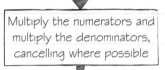

Fractions

1 Adding or subtracting fractions

| Add or subtract the whole numbers |

↓

| Write the fractions as fractions with the same denominator |

↓

| Add or subtract the fractions |

↓

| If you have an improper fraction then convert to a mixed number and add |

$2\frac{2}{3} + 1\frac{1}{2}$

$= 3 + \frac{2}{3} + \frac{1}{2}$

$= 3 + \frac{4}{6} + \frac{3}{6}$

$= 3 + \frac{7}{6}$

$= 3 + 1\frac{1}{6}$

$= 4\frac{1}{6}$

2 Dividing fractions

| Convert any mixed numbers to improper fractions |

↓

| Turn the second fraction 'upside down' and change ÷ to × |

↓

| Multiply the numerators and multiply the denominators, cancelling where possible |

↓

| Convert any improper fractions to mixed numbers |

$6\frac{1}{4} \div 1\frac{7}{8}$

$= \frac{25}{4} \div \frac{15}{8}$

$= \frac{5\cancel{25}}{1\cancel{4}} \times \frac{\cancel{8}^2}{\cancel{15}_3}$

$= \frac{10}{3}$

$= 3\frac{1}{3}$

3 Multiplying fractions

| Convert any mixed numbers to improper fractions |

↓

| Simplify if possible |

↓

| Multiply the numerators and multiply the denominators |

Converting between recurring decimals and fractions is covered on page 5.

Fractions and decimals

To convert a fraction into a decimal you divide the numerator by the denominator. Remember these common fraction-to-decimal conversions:

$\frac{1}{100} = 0.01$ $\frac{1}{10} = 0.1$ $\frac{1}{2} = 0.5$

$\frac{1}{5} = 0.2$ $\frac{1}{4} = 0.25$ $\frac{3}{4} = 0.75$

Worked example

grade C

Work out $3\frac{1}{4} \times 2\frac{2}{3}$

Give your answer in its simplest form.

$3\frac{1}{4} \times 2\frac{2}{3} = \frac{13}{4} \times \frac{8}{3}$

$= \frac{13 \times \cancel{8}^2}{1\cancel{4} \times 3}$

$= \frac{26}{3}$

$= 8\frac{2}{3}$

EXAM ALERT!

Over 40% of students got 0 marks for this question. Do **not** multiply the whole numbers and fractions separately. Always start by converting each mixed number into an improper fraction.

Do simplify calculations by 'cancelling' if possible. Then multiply the numerators **and** the denominators. Give your final answer as a mixed number.

This was a real exam question that caught students out – **be prepared!**

ResultsPlus

Now try this

edexcel

(a) A machine tool is made from two parts.
One part has a length of $1\frac{3}{4}$ inches.
The other part has a length of $2\frac{2}{3}$ inches.
What is the total length, in inches, of the
machine tool?   **(3 marks)**

(b) $3\frac{3}{4}$ is bigger than $1\frac{19}{21}$
How many times bigger?   **(3 marks)**

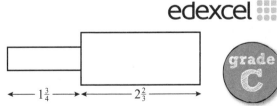

$\longleftarrow 1\frac{3}{4} \longrightarrow \longleftarrow 2\frac{2}{3} \longrightarrow$

Diagram **NOT**
accurately drawn

grade C

A*
A
B
C
D

Decimals

You can write a terminating decimal as a fraction with denominator 10, 100 or 1000.

$0.24 = \frac{24}{100} = \frac{6}{25}$ Simplify your fraction as much as possible.

To convert a fraction into a decimal you divide the numerator by the denominator.

$\frac{2}{5} = 2 \div 5 = 0.4$

It's useful to remember these common fraction-to-decimal conversions:

Fraction	$\frac{1}{100}$	$\frac{1}{10}$	$\frac{1}{2}$	$\frac{1}{5}$	$\frac{1}{4}$	$\frac{3}{4}$
Decimal	0.01	0.1	0.5	0.2	0.25	0.75

Recurring vs terminating

A terminating decimal can be written exactly as a decimal.

Recurring decimals have one digit or group of digits repeated forever. You can use dots to show the recurring digit or group of digits.

$\frac{2}{3} = 0.6666... = 0.\dot{6}$ The dot tells you that the 6 repeats forever.

$\frac{346}{555} = 0.623\,4234... = 0.6\dot{2}3\dot{4}$
These dots tell you that the group of digits 234 repeats forever.

To check whether a fraction produces a recurring decimal or a terminating decimal, write it in its simplest form and find the prime factors of its denominator.
- Prime factors only 2 and 5
  → terminating decimal
- Prime factors other than 2 or 5
  → recurring decimal

$\frac{3}{20} = \frac{3}{2^2 \times 5}$ → terminating

$\frac{5}{24} = \frac{5}{2^3 \times 3}$ → recurring

Worked example

grade
D

Using the information that
$58 \times 71 = 4118$
write down the value of
(a) 58×0.71

41.18

(b) 5800×7.1

41 180

This question would appear on your **non-calculator** paper. You should use the information given in the question to save time.

(a) 71 has been divided by 100 and 58 hasn't been changed. So the answer needs to be divided by 100:
$4118 \div 100 = 41.18$

(b) 58 has been multiplied by 100 and 71 has been divided by 10.

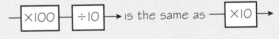
×100 — ÷10 → is the same as — ×10 →

The answer needs to be multiplied by 10:
$4118 \times 10 = 41\,180$

Now try this

edexcel

grade
D

1. Using the information that $23 \times 65 = 1495$ write down the value of
(a) 0.23×65
(b) 230×0.65 **(2 marks)**

2. $97 \times 123 = 11\,931$
Use this information to write down the answer to:
(a) 9.7×12.3
(b) $0.97 \times 123\,000$
(c) $11.931 \div 9.7$ **(3 marks)**

grade
C

Recurring decimals

Some fractions produce terminating decimals and some produce recurring decimals. You need to use algebra to convert a recurring decimal back into a fraction.

Look at page 4 for the definitions of terminating and recurring decimals.

If the question says PROVE, you MUST use algebra and write down all the steps in your working.

Worked example

grade A

Prove that $0.\dot{2}\dot{4} = \frac{8}{33}$

Let $n = \quad 0.242\,424\,24\ldots$
$100n = 24.242\,424\,24\ldots$
$\underline{\quad - n = \quad 0.242\,424\,24\ldots}$

$99n = 24$
$\quad n = \dfrac{24}{99} = \dfrac{8}{33}$

Some calculators will convert recurring decimals into fractions for you. But the question says **prove** so you must write down all the steps shown here.

1. Write the recurring decimal equal to n, and write out some of its digits.
2. Multiply both sides by 100 as there are 2 recurring digits.
3. Subtract n to remove the recurring part.
4. Divide both sides by 99 to write n as a fraction.
5. Simplify the fraction.

Multiply by…

10 if 1 digit recurs. ✓
100 if 2 digits recur. ✓
1000 if 3 digits recur. ✓

In this recurring decimal the digit 4 does not recur. Follow the same steps to write n as a fraction. After you divide by 99, multiply the top and bottom of your fraction by 10 to convert the decimal in the numerator into an integer.

Worked example

grade A

Prove that $0.4\dot{7}\dot{3}$ can be written as the fraction $\frac{469}{990}$

Let $n = \quad 0.473\,737\,37\ldots$
$100n = 47.373\,737\,37\ldots$
$\underline{\quad - n = \quad 0.473\,737\,37\ldots}$

$99n = 46.9$
$\quad n = \dfrac{46.9}{99}$
$\quad n = \dfrac{469}{990}$

Now try this

grade A

edexcel ⠿

1. Prove that $0.\dot{1}\dot{8}$ can be written as the fraction $\frac{18}{99}$

   (3 marks)

   You might be able to use your answer to part (a) to speed up your working for part (b).

2. (a) Convert the recurring decimal $0.\dot{3}\dot{6}$ to a fraction. **(3 marks)**

   grade A

   (b) Convert the recurring decimal $2.1\dot{3}\dot{6}$ to a mixed number. Give your answer in its simplest form. **(2 marks)**

-A*-
-A-
-B-
-C-
-D-

Rounding and estimation

1 To ROUND a number, you look at the next digit on the right.
5 or more → round up   less than 5 → round down

2 Decimals can be rounded to a given number of DECIMAL PLACES.
6.475 = 6.48 correct to 2 decimal places

3 To write a number correct to 3 SIGNIFICANT FIGURES (3 s.f.), look at the fourth significant figure.
0.003 07$\underline{9}$ = 0.003 08 to 3 s.f.

4 Leading zeros in decimals are not counted as significant.

5 Remember that the rule for significant figures still applies to WHOLE NUMBERS.
27 = 30 to 1 s.f.

Estimating answers

You can estimate the answer to a calculation by rounding each number to 1 SIGNIFICANT FIGURE and then doing the calculation.

This is useful for checking your answers.

$4.32 \times 18.09 \approx 4 \times 20$

'≈' means 'is approximately equal to'.

The calculation is approximately equal to 80.

Decimal division trick

You might have to divide by a decimal on your non-calculator paper. If you multiply both numbers in a division by the same amount the answer stays the same.

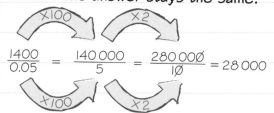

$$\frac{1400}{0.05} = \frac{140\,000}{5} = \frac{280\,00\cancel{0}}{1\cancel{0}} = 28\,000$$

Worked example

grade C

Work out an estimate for the value of

(a) $\dfrac{7.82 \times 620}{0.525}$

$$\frac{7.82 \times 620}{0.525} \approx \frac{8 \times 600}{0.5} = \frac{4800}{0.5} = 9600$$

(b) $\dfrac{6.8 \times 191}{0.051}$

$$\frac{6.8 \times 191}{0.051} \approx \frac{7 \times 200}{0.05} = \frac{1400}{0.05} = 28\,000$$

EXAM ALERT!

Only 1 student in 6 got all 3 marks for part (b). Do **not** automatically round to the nearest whole number. To give a full answer you need to:

1. Round all the numbers to **1 significant figure**.

2. Write down the calculation with rounded values.

3. Work out the answer using the correct order of operations.

4. Show every step of your working.

This was a real exam question that caught students out – **be prepared!**   Results**Plus**

Now try this

grade C

1. Work out an estimate for $\dfrac{412 \times 5.904}{0.195}$

(3 marks)

2. Work out an estimate for the value of

$$\frac{637}{3.2 \times 9.8}$$

(3 marks)

grade C

edexcel

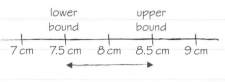

Upper and lower bounds

A* A B C D

Upper and lower bounds are a measure of accuracy.

For example, the width of a postcard is given as 8 cm to the nearest cm.

lower bound   upper bound

7 cm   7.5 cm   8 cm   8.5 cm   9 cm

The actual width of the postcard could be anything between 7.5 cm and 8.5 cm.

7.5 cm is called the LOWER BOUND or MINIMUM VALUE the width could be.

8.5 cm is called the UPPER BOUND or MAXIMUM VALUE the width could be.

Using upper and lower bounds in calculations

To find the greatest possible and least possible values of a calculation use these rules.

	+	−	×	÷
Greatest value	UB + UB	UB − LB	UB × UB	UB ÷ LB
Least value	LB + LB	LB − UB	LB × LB	LB ÷ UB

Least value of $a + b$ = lower bound of a + lower bound of b

Worked example

grade A

Sarah uses this formula to work out her average speed for a marathon.

Average speed $= \dfrac{d}{t}$

$d = 26.2$ miles correct to 3 significant figures
$t = 3.1$ hours correct to 1 decimal place

Calculate the <u>least possible value</u> for Sarah's average speed.

Give your answer correct to 2 decimal places.

	Lower bound	Upper bound
d (miles)	26.15	26.25
t (hours)	3.05	3.15

Least possible value $= \dfrac{\text{lower bound of } d}{\text{upper bound of } t}$

$= \dfrac{26.15}{3.15} = 8.301\,587... = 8.30$ (to 2 d.p.)

The least possible value is 8.30 mph (to 2 d.p.)

The question asks you to find the least possible value for the speed. This tells you that you need to use upper and lower bounds.

Write down all the figures on your calculator display before writing your rounded answer.

Degree of accuracy

If you're going for an A* you might need to give a value to 'an appropriate degree of accuracy'.

upper bound of $x = 1.17956$
lower bound of $x = 1.17892$

2 decimal places would be an appropriate degree of accuracy for x. The upper and lower bounds both round to 1.18 to 2 decimal places.

Now try this

edexcel

Katy drove for 238 miles, correct to the nearest mile. She used 27.3 litres of petrol, to the nearest tenth of a litre.

Petrol consumption $= \dfrac{\text{number of miles travelled}}{\text{number of litres of petrol used}}$

Work out the upper bound for the petrol consumption for Katy's journey.

Give your answer correct to 2 decimal places.   **(3 marks)**

grade A

A*
A
B
C
D

Fractions and percentages

'PER CENT' means 'OUT OF 100'.
You can write a percentage as a fraction with denominator 100. $73\% = \dfrac{73}{100}$ ← Numerator ← Denominator

1 To find a fraction of an amount:

| Divide by the denominator |

↓

| Multiply by the numerator |

$\dfrac{2}{5}$ of 90 kg:

  90 kg ÷ 5 = 18 kg
  18 kg × 2 = 36 kg

2 To write one quantity as a percentage of another:

| Divide the first quantity by the second quantity |

↓

| Multiply by 100 |

7 out of 20:

  $\dfrac{7}{20} \times 100 = 35\%$

3 To find a percentage of an amount:

| Divide the percentage by 100 |

↓

| Multiply by the amount |

45% of 200 g:

  $\dfrac{45}{100} \times 200\,g = 90\,g$

Worked example

 grade D

In a year group of 96 students,
60 owned a bicycle.
Express 60 as a percentage of 96

60 ÷ 96 = 0.625
0.625 × 100 = 62.5
62.5% of the students owned a bicycle.

Non-calculator: $\dfrac{60}{96} = \dfrac{5}{8} = 0.625$
                              $= 62.5\%$

Check it!
62.5% of 96 $= \dfrac{62.5}{100} \times 96 = 60$ ✓

Worked example

 grade C

120 students each chose one PE activity.
$\dfrac{1}{5}$ of the students chose swimming.
$\dfrac{3}{8}$ of the students chose tennis.
All the rest of these students chose cricket.
How many students chose cricket?

$\dfrac{1}{5}$ of 120:   120 ÷ 5 = 24

$\dfrac{3}{8}$ of 120:   120 ÷ 8 = 15; 15 × 3 = 45

120 – 24 – 45 = 51
51 students chose cricket.

Non-calculator methods

You might have to calculate percentages on your NON-CALCULATOR paper. You can use multiples of 1% and 10% to calculate percentages without a calculator.

Work out 12.5% of £600.

10% of £600 is £60   600 ÷ 10 = 60
1% of £600 is £6   600 ÷ 100 = 6
0.5% of £600 is £3   6 ÷ 2 = 3

So, 12.5% of £600 is
£60 + £6 + £6 + £3 = £75

Now try this

 edexcel

A choir has 56 members. 35 of the members are female.

(a) Work out 35 out of 56 as a percentage.   **(2 marks)**

40% of the 35 female members wear glasses.

(b) Write the number of female members who wear glasses as a fraction of the
     56 members. Give your answer in its simplest form.   **(2 marks)**

grade D

Percentage change

A*
A
B
C
D

There are two methods that can be used to increase or decrease an amount by a percentage.

£280

25% OFF

Method 1

Work out 25% of £280:

$$\frac{25}{100} \times £280 = £70$$

Subtract the decrease:

£280 − £70 = £210

Method 2

New price is 100% − 25% = 75% of original price

Multiplier:

$$75\% = \frac{75}{100} = 0.75$$

0.75 × £280 = £210

Worked example

grade D

A football club <u>increases</u> the prices of its season tickets by 4.8% each year.

In 2010 a top-price season ticket cost £550 Calculate the price of this season ticket in 2011.

$$\frac{4.8}{100} \times £550 = £26.40$$

£550 + £26.40 = £576.40

When working with money, answers must be given to 2 decimal places.
If you use Method 2, your multiplier would be 1.048

Increase or decrease?

Words meaning a decrease:
    Reduces, Depreciates, Sale

Words meaning an increase:
    Plus VAT, Interest

Calculating a percentage increase or decrease

Work out the amount of the increase or decrease

Was £60
Now £39

60 − 39 = 21

$$\frac{21}{60} = 35\%$$

This is a 35% decrease.

Write this as a percentage of the **original** amount

For a reminder about writing one quantity as a percentage of another, have a look at page 8.

A question may ask you to calculate a percentage **profit** or **loss** rather than an increase or decrease.

Now try this

grade D

1. The normal cost of a coat is £94

   In a sale the cost of the coat is reduced by 36%.

   Work out the sale price of the coat. **(3 marks)**

2. Ron went to Spain.
   He changed pounds (£) into euros (€).
   The exchange rate was £1 = €1.40
   The value of the pound has decreased from €1.40 to €1.33
   Calculate the percentage decrease in the value of the pound. **(3 marks)**

   grade C

A*
A
B
C
D

Reverse percentages and compound interest

In some questions you are given an amount after a percentage change.
To find the original amount you DIVIDE by the MULTIPLIER.

Worked example

 grade B

In a sale, normal prices are reduced by 12%.
The sale price of a digital camera is £132.88
Work out the normal price of the digital camera.

$1 - \frac{12}{100} = 0.88$

$132.88 \div 0.88 = 151$

The original price was £151.

EXAM ALERT!

Nearly 70% of students scored 0 marks on this question. Do **not** increase the price given by 12%.

The multiplier for a 12% decrease is 0.88. Divide by the multiplier to find the normal price.

Check it!
Reduce £151 by 12%.
£151 × 0.88 = £132.88 ✓

This was a real exam question that caught students out – **be prepared!**   Results **Plus**

Compound interest

You need to remember this formula:

Amount after n years
= (starting amount) × (multiplier)^n

Worked example

grade B

Raj invested £4500 for 3 years in a savings account.
He was paid 4% <u>compound interest</u> per year.
How much did Raj have in his savings account after 3 years?

$4500 \times (1.04)^3 = 5061.888$
Raj had £5061.89 in his savings account.

Read compound interest questions very carefully. Are you asked to work out the amount of **interest** or the **final amount** in the account?

Use the formula for compound interest.
Starting amount = £4500;
multiplier = 100% + 4% = 104% = 1.04;
n = 3 years.

If you see the words 'compound interest' in a question, you can work out the interest for each year separately. This is useful if you can't remember the formula.

After 3 years Raj had £5061.89

Year	Starting balance (£)	Amount of interest (£)
1	4500	$\frac{4}{100} \times 4500 = 180$
2	4500 + 180 = 4680	$\frac{4}{100} \times 4680 = 187.20$
3	4680 + 187.20 = 4867.20	$\frac{4}{100} \times 4867.20 = 194.688$

Now try this

grade B

edexcel

1. Toby invested £4500 for 2 years in a savings account.
   He was paid 4% per annum compound interest.
   How much did Toby have in his savings account after 2 years?   **(3 marks)**

2. Jaspir invested £2400 for n years in a savings account.
   He was paid 7.5% per annum compound interest.
   At the end of the n years he had £3445.51 in the savings account.
   Work out the value of n.   **(2 marks)**   grade A

Ratio

Ratios are used to compare quantities. You can find EQUIVALENT RATIOS by multiplying or dividing by the same number.

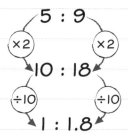

This equivalent ratio is in the form $1 : n$. This is useful for calculations.

Simplest form

To write a ratio in its simplest form, find an equivalent ratio with the smallest possible whole number values.

Simplest form
$5 : 1$ $10 : 9$
$2 : 3 : 4$

NOT simplest form
$1 : 1.5 : 2$ $10 : 2$
$1 : 0.9$

Worked example
grade **C**

Three women won a total of £360
They shared the money in the ratio $7 : 3 : 2$
Donna received the largest amount.
Work out the amount Donna received.

$7 + 3 + 2 = 12$
$\dfrac{£360}{12} = £30$
$£30 \times 7 = £210$
Donna received £210.

1. Work out the total number of parts.
2. Work out the value of 1 part.
3. Work out 7 parts (largest part) for Donna.

Worked example
grade **C**

Jared is preparing the ingredients for a pizza.
He uses cheese, topping and dough in the ratio $2 : 3 : 5$
He uses 70 grams of dough.
How many grams of topping will Jared use?

$\dfrac{70}{5} = 14$
$3 \times 14\,g = 42\,g$ of topping

cheese : topping : dough
$2 : 3 : 5$
$? : ? : 70\,g$
5 parts = 70 g so 1 part = 14 g
Multiply by 3 to find the weight of topping used.

Now try this
grade **C**

1. Alex and Ben were given a total of £360
They shared the money in the ratio $5 : 7$
Ben received more money than Alex.
How much more? **(2 marks)**

First convert 0.5 kg into grams. Then work out how much crumble topping Tom needs to make.
Check it!
weight of sugar + butter + flour
= weight of crumble topping

edexcel

2. Tom is making plum crumble.
For the crumble topping, he uses sugar, butter and flour in the ratio $2 : 3 : 5$
grade **C**
Tom uses 250 g of plums for every 400 g of crumble topping.
Tom has 0.5 kg of plums. He uses all of them.
Work out how much sugar, butter and flour he will need for the crumble topping. **(4 marks)**

A*
A
B
C
D

Proportion

This page introduces direct proportion and inverse proportion.
Direct proportion is covered in more detail (up to grade A) on page 43.
Inverse proportion is covered in more detail (up to grade A) on page 44.

Two quantities are in DIRECT PROPORTION when both quantities increase at the same rate.

Number of theatre tickets bought   Total cost

3   £135
×3   ×3
9   £405

Two quantities are in INVERSE PROPORTION when one quantity increases at the same rate as the other quantity decreases.

Average speed   Time taken

40 km/h   2 hours
×2   ÷2
80 km/h   1 hour

Worked example

grade D

Rujuta buys 2 cupcakes for a total cost of £3.30
Work out the cost of 7 of these cupcakes.

Cost of 1 cupcake $= \dfrac{£3.30}{2} = £1.65$

Cost of 7 cupcakes $= £1.65 × 7 = £11.55$

Calculate the cost of 1 cupcake first. Then multiply the cost of 1 cupcake by 7 to work out the cost of 7 cupcakes. Remember to write down the correct units. When working with money, you must give your answer to 2 decimal places.

Divide or multiply?

6 people can build a wall in 4 days.
How long would it take 8 people to build the same wall?
Inverse proportion problems often involve time. The more people working on a task, the quicker it will be finished.

You can solve this problem by working out how long it would take 1 person to build the wall. Use common sense to decide whether to divide or multiply.

6 × 4 = 24 so 1 person could build the wall in 24 days.
You multiply because it would take 1 person more time to build the wall.

24 ÷ 8 = 3 so 8 people could build the wall in 3 days.
You divide because it would take 8 people less time to build the wall.

Now try this

If the speed is less, the mass will be bigger.

edexcel

1. Amy bought 17 footballs for a total cost of £50.83

grade D

James bought 11 footballs.

The cost of each football bought by Amy and James was the same.

Work out how much James paid for his 11 footballs.   **(2 marks)**

2. A machine can project balls to help with coaching. The velocity, v km/h, at which the ball is projected is inversely proportional to the mass, m grams, of the ball. A ball with a mass of 150 g is projected at a speed of 48 km/h. Work out the mass of a ball projected at 24 km/h.   **(3 marks)**

grade C

Indices 2

A*
A
B
C
D

1 Negative powers

$$a^{-n} = \frac{1}{a^n}$$

$$5^{-2} = \frac{1}{5^2} = \frac{1}{25}$$

Be careful!

A NEGATIVE power can still have a POSITIVE answer.

2 Reciprocals

$$a^{-1} = \frac{1}{a}$$

This means that a^{-1} is the RECIPROCAL of a.

You can find the reciprocal of a fraction by turning it upside down.

$$\left(\frac{5}{9}\right)^{-1} = \frac{9}{5}$$

3 Powers of fractions

$$\left(\frac{a}{b}\right)^n = \frac{a^n}{b^n}$$

$$\left(\frac{3}{10}\right)^2 = \frac{3^2}{10^2} = \frac{9}{100}$$

4 Combining rules

You can apply the rules one at a time.

$$\left(\frac{a}{b}\right)^{-n} = \frac{b^n}{a^n}$$

Turn the fraction upside down and then raise the numerator and denominator to the given power.

$$\left(\frac{2}{3}\right)^{-3} = \left(\frac{3}{2}\right)^3 = \frac{3^3}{2^3} = \frac{27}{8}$$

5 Fractional powers

You can use fractional powers to represent roots.

$$a^{\frac{1}{2}} = \sqrt{a} \qquad 49^{\frac{1}{2}} = \pm 7$$

$$a^{\frac{1}{3}} = \sqrt[3]{a} \qquad 27^{\frac{1}{3}} = 3$$

$$a^{\frac{1}{4}} = \sqrt[4]{a} \qquad 16^{\frac{1}{4}} = \pm 2$$

CHECK IT!

A whole number raised to a power less than 1 gets smaller.

6 More complicated indices

You can use the index laws to work out more complicated fractional powers.

$$a^{\frac{m}{n}} = \left(a^{\frac{1}{n}}\right)^m$$

Do these calculations ONE STEP AT A TIME.

$$27^{-\frac{2}{3}} = (27^{\frac{1}{3}})^{-2}$$
$$= \left(\sqrt[3]{27}\right)^{-2}$$
$$= 3^{-2} = \frac{1}{3^2}$$
$$= \frac{1}{9}$$

3^n is not the same as $3n$. You can't divide by 3 to get n on its own.

Instead of trying to get n on its own, try to make the base on the right-hand side the same as the base on the left-hand side.

1. Write 9 as a power of 3. Remember to use brackets.
2. Use $(a^n)^m = a^{nm}$ to write the right-hand side as a single power of 3.
3. Compare both sides and write down the value of n.

Worked example

grade A*

Find the value of n when $3^n = 9^{-\frac{3}{2}}$

$$3^n = (3^2)^{-\frac{3}{2}}$$

$3^n = 3^{-3}$ because $2 \times -\frac{3}{2} = -3$

So $n = -3$

Now try this

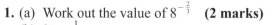
grade A   edexcel

1. (a) Work out the value of $8^{-\frac{2}{3}}$ **(2 marks)**
   (b) $3^n = \frac{1}{9}$
       Find the value of n. **(2 marks)**

grade A

2. (a) Work out the value of
      (i) 3^{-2} (ii) $36^{\frac{1}{2}}$
      (iii) $27^{\frac{2}{3}}$ (iv) $\left(\frac{16}{81}\right)^{-\frac{3}{4}}$ **(5 marks)**
   (b) $4n^{\frac{3}{2}} = 8^{-\frac{1}{3}}$
       Find the value of n. **(3 marks)**

grade A*

A* A B C D

Standard form

Numbers in standard form have two parts.

$$7.3 \times 10^{-6}$$

This part is a number greater than or equal to 1 and less than 10

This part is a power of 10

You can use standard form to write very large or very small numbers.

$$920\,000 = 9.2 \times 10^5$$

Numbers greater than 10 have a positive power of 10

$$0.00703 = 7.03 \times 10^{-3}$$

Numbers less than 1 have a negative power of 10

Counting decimal places

You can count decimal places to convert between numbers in standard form and ordinary numbers.

3 jumps

$7\,900 = 7.9 \times 10^3$

7900 > 10
So the power is positive

4 jumps

$0.00035 = 3.5 \times 10^{-4}$

0.00035 < 1
So the power is negative

BE CAREFUL!

Don't just count zeros to work out the power.

Multiplying numbers in standard form

| Rearrange so powers of 10 are together |
| Multiply the number parts |
| Add the powers |
| Rewrite your answer in standard form if necessary |

$(3 \times 10^3) \times (5 \times 10^6)$
$= (3 \times 5) \times (10^3 \times 10^6)$
$= 15 \times 10^9$

$a^m \times a^n = a^{m+n}$

$= 1.5 \times 10^1 \times 10^9$
$= 1.5 \times 10^{10}$

Dividing numbers in standard form

| Rearrange so powers of 10 are together |
| Divide the number parts |
| Subtract the powers |
| Rewrite your answer in standard form if necessary |

$(1.2 \times 10^8) \div (2 \times 10^4)$
$= (1.2 \div 2) \times (10^8 \div 10^4)$
$= 0.6 \times 10^4$

$a^m \div a^n = a^{m-n}$

$= 6 \times 10^{-1} \times 10^4$
$= 6 \times 10^3$

Worked example

grade B

(a) Write 82 500 000 in standard form.
$82\,500\,000 = 8.25 \times 10^7$

(b) Write 6.5×10^{-4} as an ordinary number.
$6.5 \times 10^{-4} = 0.000\,65$

Be careful!

To add or subtract numbers in standard form, first change them into ordinary numbers.

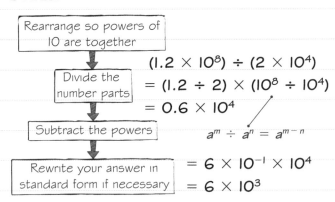

EXAM ALERT!

Only half of students got the mark for part (a).

Make sure you count the number of decimal places you need to move to get a number between 1 and 10.

82 500 000 is bigger than 10 so the power will be positive.

7 jumps

$8\,2\,5\,0\,0\,0\,0\,0$

This was a real exam question that caught students out – **be prepared!** Results Plus

Now try this

$A = 6 \times 10^2 \quad B = 8 \times 10^4$

(a) Work out A + B.
   Give your answer in standard form.

grade B

(2 marks)

$C = 2 \times 10^5 \quad D = 3 \times 10^4$

(b) Work out CD.
   Give your answer in standard form.

grade A

(3 marks)

 edexcel

14

Calculator skills

These calculator keys are really useful.

x^2 Square a number.

$(-)$ Enter a negative number.

x^3 Cube a number.

$\sqrt{\square}$ Find the square root of a number.

x^{-1} Find the reciprocal of a number.

$\sqrt[3]{\square}$ Find the cube root of a number. You might need to press the shift key first.

Ans Use your previous answer in a calculation.

S⇔D Change the answer from a fraction or surd to a decimal. Not all calculators have this key.

Standard form

You can enter numbers in standard form using the $\boxed{\times 10^x}$ key.

To enter 3.7×10^{-6} press

3 . 7 ×10ˣ (−) 6

For a reminder about how to convert between decimal numbers and standard form have a look at page 14.

Calculator checklist

Use the same calculator in your exam as you have used during your course. ✓

Make sure you know how to use all these functions on your calculator. ✓

Show your working even if you use a calculator to answer a question. ✓

Write down all the figures on your calculator display then round your answer. ✓

If $\sin 30° = \frac{1}{2}$ then you know your calculator is in the right mode. ✓

Worked example

grade C

Use your calculator to work out the value of

$$\frac{(7.91 - \sqrt[3]{81}) \times 4.32}{6.23 + 1.491}$$

Write down all the figures on your calculator display.

$$\frac{(7.91 - \sqrt[3]{81}) \times 4.32}{6.23 + 1.491} = \frac{3.583\,25... \times 4.32}{7.721}$$

$$= \frac{15.479\,64...}{7.721}$$

$$= 2.004\,875\,737$$

Calculate $\sqrt[3]{81}$ using the $\boxed{\sqrt[3]{\square}}$ key.
Always show what the top of the fraction comes to as well as the bottom.

Remember to write down **all** the figures on your calculator display.

Check it!
Do the whole calculation in one go on your calculator, using brackets and the $\boxed{\square}$ key.

Now try this

grade B

1. Work out $\dfrac{\sqrt{2.56} + \sin 57°}{8.765 - 6.78}$

   Write down all the figures on your calculator display.   **(3 marks)**

edexcel

2. Work out $\dfrac{2 \times 4.5 \times 10^9 \times 1.8 \times 10^9}{2.5 \times 10^9 - 1.8 \times 10^9}$

   Give your answer in standard form correct to 3 significant figures.   **(3 marks)**

grade A

A* A B C D

Surds

You can give exact answers to calculations by leaving some numbers as square roots.

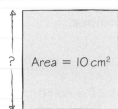

? | Area = 10 cm²

This square has a side length of $\sqrt{10}$ cm.

You can't write $\sqrt{10}$ exactly as a decimal number. It is called a SURD.

Rules for simplifying square roots

These are the most important rules to remember when dealing with surds.

$$\sqrt{ab} = \sqrt{a} \times \sqrt{b} \qquad \sqrt{8} = \sqrt{4} \times \sqrt{2} = 2\sqrt{2}$$

$$\sqrt{\frac{a}{b}} = \frac{\sqrt{a}}{\sqrt{b}} \qquad \sqrt{\frac{3}{25}} = \frac{\sqrt{3}}{\sqrt{25}} = \frac{\sqrt{3}}{5}$$

You need to remember these rules for your exam. They are NOT given on the formula sheet.

Worked example

grade A

Write $\sqrt{45}$ in the form $k\sqrt{5}$ where k is an integer.

$$\sqrt{45} = \sqrt{9 \times 5}$$
$$= \sqrt{9} \times \sqrt{5}$$
$$= 3\sqrt{5}$$
$$k = 3$$

Do **not** leave an answer as a square root if it can be written as an integer.
1. Look for a factor of 45 which is a square number: $45 = 9 \times 5$.
2. Use the rule $\sqrt{ab} = \sqrt{a} \times \sqrt{b}$ to split the square root into two square roots.
3. Write $\sqrt{9}$ as a whole number.

RATIONALISING THE DENOMINATOR of a fraction means making the denominator a whole number.

You can do this by multiplying the top AND bottom of the fraction by the surd part in the denominator.

$$\frac{5}{3\sqrt{2}} = \frac{5\sqrt{2}}{6}$$

The surd part of the denominator is $\sqrt{2}$

Remember that $\sqrt{2} \times \sqrt{2} = 2$
So $3\sqrt{2} \times \sqrt{2} = 3 \times 2 = 6$

Good form

Most surd questions ask you to write a number or answer in a certain FORM.

This means you need to find INTEGERS for all the letters in the expression.

$6\sqrt{3}$ is in the form $k\sqrt{3}$.
$$k = 6$$
The integers can be positive or negative.
$4 - 9\sqrt{2}$ is in the form $p + q\sqrt{2}$.
$$p = 4 \text{ and } q = -9$$
You can check your answer by writing down the integer value for each letter.

Now try this

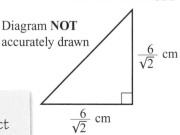

edexcel

grade A

(a) Express $\frac{6}{\sqrt{2}}$ in the form $a\sqrt{b}$, where a and b are positive integers. **(2 marks)**

The diagram shows a right-angled isosceles triangle.

The length of each of its equal sides is $\frac{6}{\sqrt{2}}$ cm.

grade A*

(b) Find the area of the triangle. Give your answer as an integer. **(2 marks)**

Remember to use the correct formula for the area of a triangle.

Diagram **NOT** accurately drawn

$\frac{6}{\sqrt{2}}$ cm

$\frac{6}{\sqrt{2}}$ cm

Problem-solving practice

About half of the questions on your exam will need problem-solving skills.

These skills are sometimes called AO2 and AO3.

Practise using the questions on the next two pages.

For these questions you might need to:

• choose which mathematical technique or skill to use

• apply a technique in a new context

• plan your strategy to solve a longer problem

• show your working clearly and give reasons for your answers.

 Susan has 2 dogs.

Each dog is fed $\frac{3}{8}$ kg of dog food each day.

Susan buys dog food in bags.

Each bag weighs 14 kg.

For how many days can Susan feed the 2 dogs from 1 bag of dog food?

You must show ALL your working. (5 marks)

Fractions p. 3

There are lots of steps in this question so make sure you keep track of your working.

TOP TIP

If your working is very untidy or hard to follow, then re-write it clearly and cross out your original working.

---

 *Amy has a field in the shape of a trapezium.

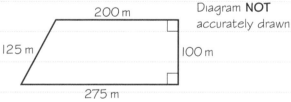

Diagram **NOT** accurately drawn

She wants to sell the field.

Farmer Boyce offers her £1 per m².

Farmer Giles offers her £24 000

Which is the better offer? (4 marks)

Ratio p. 11
Perimeter and area p. 55

You need to find the area of the field first. You can use the formula for the area of a trapezium or divide the field into a rectangle and a triangle. You should work out how much Farmer Boyce is offering for the field and then write down who has made the better offer.

TOP TIP

If a question has a * next to it, then there are marks available for QUALITY OF WRITTEN COMMUNICATION. This means you must show all your working and explain why you have chosen a particular farmer.

Problem-solving practice

3 This item appeared in a newspaper.

> **Cow produces 3% more milk**
> A farmer found that when his cow listened to classical music the milk it produced increased by 3%.
> This increase of 3% represented 0.72 litres of milk.

Calculate the amount of milk produced by the cow when it listened to classical music. **(3 marks)**

Proportion p. 12
Percentage change p. 9

grade **B**

When the cow listened to classical music, it produced 103% of the milk it produced originally. You know that 3% represents 0.72 litres. Use this information to work out what 103% represents.

TOP TIP

You can sometimes solve percentage problems by working out what 1% represents.

---

4 Aminata invested £2500 for n years in a savings account.

She was paid 3% per annum compound interest.

At the end of n years, Aminata had £2813.77 in the savings account.

Work out the value of n. **(2 marks)**

Reverse percentages and compound interest p. 10

grade **A**

You are normally given the number of years for a compound interest question and have to calculate the amount. Here you need to work out the number of years, n. You know that n is going to be a whole number, so work out the totals after 1 year, 2 years, 3 years, etc.

TOP TIP

Read the question carefully and make sure you know what you are being asked to work out.

5

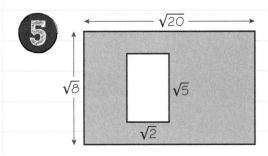

Diagram **NOT** accurately drawn

A large rectangular piece of card is $\sqrt{20}$ cm long and $\sqrt{8}$ cm wide.

A small rectangle $\sqrt{2}$ cm long and $\sqrt{5}$ cm wide is cut out of the piece of card.

Express the area of the card that is left as a percentage of the area of the large rectangle. **(4 marks)**

Surds p. 16

grade **A***

Work out an expression for the area of the large rectangle and simplify your answer as much as possible. Then work out an expression for the area of the small rectangle.

The lengths are given as surds so keep your working in surd form. **Don't** use a calculator or round your answers. Even though it doesn't say so in the question, you need to show **all your working** to give a full answer.

TOP TIP

Practise surds **without a calculator**. A question like this could appear on your non-calculator paper.

Algebraic expressions

A*
A
B
C
D

If you're going for a grade B or better you need to make sure you can work with algebraic expressions confidently.

1 You can use the INDEX LAWS to simplify algebraic expressions.

$a^m \times a^n = a^{m+n}$
$x^4 \times x^3 = x^{4+3} = x^7$

$\dfrac{a^m}{a^n} = a^{m-n}$
$m^8 \div m^2 = m^{8-2} = m^6$

$(a^m)^n = a^{mn}$
$(n^2)^4 = n^{2 \times 4} = n^8$

2 You can square or cube a whole expression.

$(4x^3y)^2 = (4)^2 \times (x^3)^2 \times (y)^2$
$\qquad = 16x^6y^2$

$16 = (4)^2$

$(x^3)^2 = x^{3 \times 2} = x^6$

You need to square everything inside the brackets.

Remember that if a letter appears on its own then it has the power 1

3 Algebraic expressions may also contain negative and fractional indices.

$a^{-m} = \dfrac{1}{a^m}$

$(c^2)^{-3} = c^{2 \times -3} = c^{-6} = \dfrac{1}{c^6}$

$a^{\frac{1}{n}} = \sqrt[n]{a}$

$(8p^3)^{\frac{1}{3}} = (8)^{\frac{1}{3}} \times (p^3)^{\frac{1}{3}}$
$\qquad = \sqrt[3]{8} \times p^{3 \times \frac{1}{3}}$
$\qquad = 2p$

One at a time

When you are MULTIPLYING expressions:

1. Multiply any number parts first.

2. Add the powers of each letter to work out the new power.

$6p^2q \times 3p^3q^2 = 18p^5q^3$

$6 \times 3 = 18$

$p^2 \times p^3 = p^{2+3} = p^5$

$q \times q^2 = q^{1+2} = q^3$

When you are DIVIDING expressions:

1. Divide any number parts first.

2. Subtract the powers of each letter to work out the new power.

$12 \div 3 = 4$

$b^3 \div b^2 = b^{3-2} = b$

$\dfrac{12a^5b^3}{3a^2b^2} = 4a^3b$

$a^5 \div a^2 = a^{5-2} = a^3$

Worked example

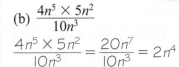

Simplify fully   grade **C**

(a) $(n^3)^3$

$(n^3)^3 = n^9$

(b) $\dfrac{4n^5 \times 5n^2}{10n^3}$

$\dfrac{4n^5 \times 5n^2}{10n^3} = \dfrac{20n^7}{10n^3} = 2n^4$ grade **B**

'Simplify' and 'simplify fully' mean you have to write each expression with a single number part and a single power of n.

(a) Use $(a^m)^n = a^{mn}$

(b) Start by simplifying the top part of the fraction. Do the number part first then the powers. Use $a^m \times a^n = a^{m+n}$
Next divide the expressions. Divide the number part then subtract the indices.
You are using $\dfrac{a^m}{a^n} = a^{m-n}$

Now try this

edexcel

grade **C**

1. Simplify
(a) $p^2 \times p^7$ **(1 mark)**
(b) $x^8 \div x^3$ **(1 mark)**
(c) $\dfrac{q^3 \times q^4 \times q}{q^2}$ **(2 marks)**

grade **B**

(d) $3s^2t^3 \times 4s^4t^2$ **(2 marks)**

2. (a) Simplify
(i) $k^5 \div k^2$ (ii) $(m^{-4})^{-2}$
(iii) $2t^2 \times 3r^3t^4$ (iv) $(3xy^2)^4$ **(6 marks)**

grade **B**

(b) Given that $\dfrac{20x^8y \times 3xy^7}{15x^3y^2} = Ax^my^n$, work out the values of A, m and n. **(3 marks)**

grade **A**

A*
A
B
C
D

Arithmetic sequences

An arithmetic sequence is a sequence of numbers where the rule is 'add a fixed number' or 'subtract a fixed number'. In your exam, you may be asked to work out the nth term of a sequence. Look at this example which shows you how to do it in four steps.

1 Here are the first five terms of an arithmetic sequence.

$$1 \;\boxed{+4}\; 5 \;\boxed{+4}\; 9 \;\boxed{+4}\; 13 \;\boxed{+4}\; 17$$

Find, in terms of n, an expression for the nth term of the sequence.

Write in the difference between each term.

2 Here are the first five terms of an arithmetic sequence.

Zero term
$$-3 \quad 1 \;\boxed{+4}\; 5 \;\boxed{+4}\; 9 \;\boxed{+4}\; 13 \;\boxed{+4}\; 17$$

Find, in terms of n, an expression for the nth term of the sequence.

Work backwards to find the **zero term** of the sequence. You need to subtract 4 from the first term.

3 Here are the first five terms of an arithmetic sequence.

Zero term
$$-3 \quad 1 \;\boxed{+4}\; 5 \;\boxed{+4}\; 9 \;\boxed{+4}\; 13 \;\boxed{+4}\; 17$$

Find, in terms of n, an expression for the nth term of the sequence.

nth term = difference × n + zero term

Write down the formula for the nth term.
Remember this formula for the exam.

4 Here are the first five terms of an arithmetic sequence.

Zero term
$$-3 \quad 1 \;\boxed{+4}\; 5 \;\boxed{+4}\; 9 \;\boxed{+4}\; 13 \;\boxed{+4}\; 17$$

Find, in terms of n, an expression for the nth term of the sequence.

nth term = difference × n + zero term
nth term = $4n - 3$

You can use the nth term to check whether a number is a term in the sequence.
The value of n in your nth term has to be a POSITIVE whole number.
So is 99 a term of the sequence?
Try some different values of n:
when $n = 25$, $4n - 3 = 4 \times 25 - 3 = 97$
when $n = 26$, $4n - 3 = 4 \times 26 - 3 = 101$
You can't use a value of n between 25 and 26 so 99 isn't a term in the sequence.
TIP!
If 99 is a term of the sequence then $4n - 3 = 99$.
You can solve this equation to show that n is not a whole number which proves that 99 is not in the sequence.

Check it!

Check your answer by substituting values of n into your nth term.
1st term: when $n = 1$,
$4n - 3 = 4 \times 1 - 3 = 1$ ✓
2nd term: when $n = 2$,
$4n - 3 = 4 \times 2 - 3 = 5$ ✓
You can also generate any term of the sequence.
For the 20th term, $n = 20$:
$4n - 3 = 4 \times 20 - 3 = 77$
So the 20th term is 77.

Now try this grade C

(a) Here are the first 5 terms of a number sequence: 3   7   11   15   19
Write down an expression, in terms of n, for the nth term of this sequence.   **(2 marks)**

(b) Adeel says that 318 is a term in the number sequence.
Is Adeel correct? You must justify your answer.   **(2 marks)**

(c) The nth term of a different number sequence is $5 - 2n$
Work out the first three terms of this number sequence.   **(3 marks)**

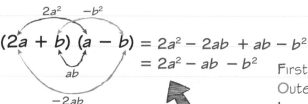

Expanding brackets

Expanding or multiplying out brackets is a key algebra skill.

You have to multiply the expression outside the bracket by everything inside the bracket.

$4n \times n^2 = 4n^3$

$$4n(n^2 + 2) = 4n^3 + 8n$$

$4n \times 2 = 8n$

'Expand and simplify' means 'multiply out and then collect like terms'.

Golden rule

When you expand, you need to be careful with negative signs in front of the bracket.

Negative signs belong to the term to their right.

$$-2 \times x \qquad -2 \times -y$$
$$x - 2(x - y) = x - 2x + 2y$$
$$= -x + 2y$$

Multiply out the brackets first and then collect like terms if possible.

You can use the GRID METHOD to expand two brackets.

$(x + 7)(x - 5) = x^2 - 5x + 7x - 35$
$\qquad\qquad\qquad = x^2 + 2x - 35$

Remember to collect like terms if possible.

	x	-5
x	x^2	$-5x$
7	$7x$	-35

The negative sign belongs to the 5. You need to write it in your grid.

OR

You can use the acronym FOIL to expand two brackets.

$$2a^2 \qquad -b^2$$
$$(2a + b)(a - b) = 2a^2 - 2ab + ab - b^2$$
$$\qquad\qquad ab$$
$$= 2a^2 - ab - b^2$$
$$-2ab$$

First terms
Outer terms
Inner terms
Last terms

Some people remember this as a 'smiley face'.

Worked example

 grade **B**

Expand and simplify $(3p - 4)^2$

$(3p - 4)^2 = (3p - 4)(3p - 4)$
$\qquad\qquad = 9p^2 - 12p - 12p + 16$
$\qquad\qquad = 9p^2 - 24p + 16$

	$3p$	-4
$3p$	$9p^2$	$-12p$
-4	$-12p$	16

The question is 'expand and simplify' so you have to multiply out **and** collect like terms.

Use the grid method or **FOIL** to find all **four** terms of the expansion.

Be extra careful with your negative signs.
$-4 \times -4 = 16 \qquad p \times -4 = -4p$

Now try this

 grade **D**

 grade **C**

 grade **B**

edexcel ⣿

1. (a) Expand $7(5 - 2x)$ **(1 mark)**

   (b) Expand and simplify
   $8(3x + 4) - 2(4x - 5)$ **(2 marks)**

   (c) Expand and simplify
   $(y - 3)(y + 4)$ **(2 marks)**

2. Expand and simplify
   $(x - 3)(2x + 5)$ **(2 marks)**

3. (a) Expand and simplify
   $(x - 5y)(2x + 3y)$ **(2 marks)**

   (b) Expand and simplify
   $(x + 6)^2 - (x - 7)^2$ **(3 marks)**

 grade **B**

A*
A
B
C
D

Factorising

Factorising is the opposite of expanding brackets:

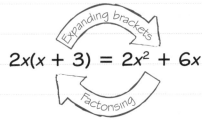

$$2x(x + 3) = 2x^2 + 6x$$

You need to look for the LARGEST FACTOR you can take out of every term in the expression.

$$10a^2 + 5ab = 5(2a^2 + ab)$$

This expression has only been PARTLY FACTORISED.

$$10a^2 + 5ab = 5a(2a + b)$$

This expression has been COMPLETELY FACTORISED.

Factorising $x^2 + bx + c$

You need to write the expression with TWO BRACKETS.

You need to find two numbers which add up to 7... $5 + 2 = 7$

$$x^2 + 7x + 10 = (x + 5)(x + 2)$$

... and multiply to make 10 $5 \times 2 = 10$

When factorising $x^2 + bx + c$, use this table to help you find the two numbers:

b	c	Factors
Positive	Positive	Both numbers positive
Positive	Negative	Bigger number positive and smaller number negative
Negative	Negative	Bigger number negative and smaller number positive
Negative	Positive	Both numbers negative

Factorising $ax^2 + bx + c$

$$2x^2 - 7x - 15 = (2x \quad)(x \quad)$$

One of the brackets must contain a $2x$ term. Try pairs of numbers which have a product of -15. Check each pair by multiplying out the brackets.

$(2x + 5)(x - 3) = 2x^2 - x - 15$ ✗
$(2x - 3)(x + 5) = 2x^2 + 7x - 15$ ✗
$(2x + 3)(x - 5) = 2x^2 - 7x - 15$ ✓

Difference of two squares

You can factorise expressions that are written as

$$(\text{something})^2 - (\text{something else})^2$$

Use this rule:
$$a^2 - b^2 = (a + b)(a - b)$$
$$x^2 - 36 = x^2 - 6^2$$
$$= (x + 6)(x - 6)$$

36 is a square number.
$36 = 6^2$ so $a = x$ and $b = 6$

Worked example

Factorise fully   grade D
(a) $x^2 + 3x$
$x(x + 3)$
(b) $x^2y + xy^2$ grade C
$xy(x + y)$

Now try this

In part (f) start with
$6f^2 - 10f + 4 = 2(3f^2 - 5f + 2)$

 edexcel

grade C

(a) Factorise completely $12x^2 - 18xy$ **(1 mark)**
(b) Factorise $a^2 - 9b^2$ **(1 mark)**
(c) Factorise $6x^2 - 9xy$ **(1 mark)**

(d) Factorise $2x^2 - 7x + 6$ **(2 marks)**
(e) Factorise $(p + q)^2 + 5(p + q)$ **(2 marks)**
(f) Factorise $6f^2 - 10f + 4$ **(2 marks)**

grade A

Linear equations 1

A*
A
B
C
D

To solve a linear equation you need to get the letter on its own on one side.
It is really important to write your working NEATLY when you are solving equations.

$$5x + 3 = 18 \quad (-3)$$

Write down the operation you are carrying out. Remember to do the same thing to both sides of the equation.

Every line of working should have an equals sign in it.

$$5x = 15 \quad (\div 5)$$
$$x = 3$$

Start a new line for each step.
Do one operation at a time.

Line up the equals signs.

Letter on both sides?

To solve an equation you have to get the letter on its own on one side of the equation.

Start by collecting like terms so that all the letters are together.

$$2 - 2x = 26 + 4x \quad (+ 2x)$$
$$2 = 26 + 6x \quad (- 26)$$
$$-24 = 6x \quad (\div 6)$$
$$-4 = x$$

You can write your answer as
$-4 = x$ or as $x = -4$

Equations with brackets

Always start by multiplying out the brackets then collecting like terms.
For a reminder about multiplying out brackets have a look at page 21.

$$19 = 8 - 2(5 - 3y)$$
$$19 = 8 - 10 + 6y$$
$$19 = -2 + 6y \quad (+ 2)$$
$$21 = 6y \quad (\div 6)$$
$$\frac{21}{6} = y$$
$$y = \frac{7}{2} \text{ or } 3\frac{1}{2} \text{ or } 3.5$$

Your answer can be written as a fraction or decimal.

Worked example

grade **D**

Solve $\quad 7r + 2 = 5(r - 4)$

$$7r + 2 = 5r - 20 \quad (- 5r)$$
$$2r + 2 = -20 \quad (- 2)$$
$$2r = -22 \quad (\div 2)$$
$$r = -11$$

Multiply out the brackets then collect all the terms in r on one side.

Check it!
Substitute $r = -11$ into each side of the equation.

Left-hand side: $7(-11) + 2 = -75$
Right-hand side: $5(-11 - 4) = -75$ ✓

Now try this

Expand the brackets first to get started.
Then move the terms in y to the side that has the highest term in y.

edexcel

1. Solve

    (a) $6x + 7 = 10$ **(2 marks)**

grade **D**

    (b) $5(x + 4) = 22$ **(2 marks)**

2. Solve

    $9(2y - 1) = 4(5y - 4)$

             (4 marks)

grade **C**

A* A B C D

Linear equations 2

Equations with fractions

When you have an equation with fractions, you need to get rid of any fractions before solving. You can do this by multiplying every term by the lowest common multiple (LCM) of the denominators.

$$\frac{x}{3} + \frac{x-1}{5} = 11 \qquad (\times 15)$$

The LCM of 3 and 5 is 15.

$$\frac{^5\cancel{15}x}{\cancel{3}_1} + \frac{^3\cancel{15}(x-1)}{\cancel{5}_1} = 165$$

Cancel the fractions. There is more about simplifying algebraic fractions on page 19.

$$5x + 3x - 3 = 165$$
$$8x - 3 = 165 \qquad (+3)$$
$$8x = 168 \qquad (\div 8)$$
$$x = 21$$

Multiplying by an expression

You might have to multiply by an expression to get rid of the fractions.

$$\frac{20}{n-3} = -5 \qquad (\times(n-3))$$
$$20 = -5(n-3)$$

Worked example

grade **C**

Solve
$$\frac{29-x}{4} = x + 5$$

$$\frac{4(29-x)}{4} = 4(x+5)$$
$$29 - x = 4(x+5)$$
$$29 - x = 4x + 20 \qquad (+x)$$
$$29 = 5x + 20 \qquad (-20)$$
$$9 = 5x \qquad (\div 5)$$
$$\frac{9}{5} = x$$

EXAM ALERT!

Less than a quarter of students got full marks on this question.

The first step is to multiply both sides of the equation by 4 to get rid of the fraction.

Use brackets to show that you are multiplying everything by 4. Do **not** write $4 \times x + 5 = 4x + 5$.

Next multiply out the brackets and then solve as normal.

Check it!

Always check every line of your working carefully.

This was a real exam question that caught students out – **be prepared!** ResultsPlus

Writing your own equations

You can find unknown values by writing and solving equations.

$4(x-1)$ cm $(3x+3)$ cm

 $\frac{5}{n}$ m

Perimeter = 20 m   2 m

$$4(x-1) = 3x + 3$$

$$\frac{5}{n} + \frac{5}{n} + 2 + 2 = 20$$

Now try this

First expand the brackets and get an equation with four terms.
Multiply **each** of these four terms by 3.

1. Solve
$$\frac{x}{3} - 5 = 3(x-2)$$
(4 marks)

grade **B**

2. Solve
$$\frac{8x}{15} + \frac{3}{5} = \frac{x+1}{2}$$
(4 marks)

grade **A**

 edexcel

 Multiply **each** of the three terms by the LCM of 15, 5 and 2.

Straight-line graphs

A straight line on a graph has the equation $y = mx + c$ where m and c are numbers.

c is the y-intercept. The line crosses the y-axis at the point $(0, c)$.

m is the gradient. This measures how steep the line is.

Worked example   grade C

(a) On the grid draw the graph of $x + y = 4$
for values of x from -2 to 5

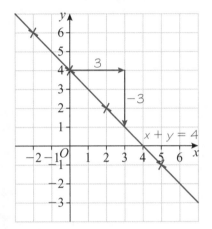

x	-2	0	2	5
y	6	4	2	-1

(b) Write down the gradient of the graph.

Gradient $= \dfrac{-3}{3} = -1$

EXAM ALERT!

Only half of students got full marks on part (a).

Use a table of values to draw a graph. Draw a triangle to find the gradient. The gradient is negative because the line slopes downwards from left to right.

This was a real exam question that caught students out – **be prepared!**   ResultsPlus

You can also identify the gradient of a straight line when its equation is written in the form $y = mx + c$.
$x + y = 4$ so $y = -x + 4$
From this equation, you can see that the gradient is -1. You can also identify the y-intercept as $(0, 4)$.

Worked example   grade A

A line passes through the points with coordinates $(1, 5)$ and $(2, 7)$.
Find the equation of the line.

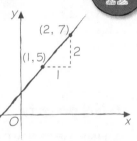

Gradient, $m = \dfrac{2}{1} = 2$
Equation of line: $y = mx + c \rightarrow y = 2x + c$
For point $(1, 5)$, $x = 1$, $y = 5$
Substitute these values in equation:
$5 = 2 + c \rightarrow c = 3$
The equation is $y = 2x + 3$.

Draw a **sketch** showing the two points. Draw a triangle to work out the gradient.
To work out the y-intercept:
1. Put your value for m into the equation of the line.
2. Use the x- and y-values of one of the points on the line to write an equation.
3. Solve your equation to find the value of c.

Now try this   edexcel   grade A

A straight line passes through the points with coordinates $(0, 17)$ and $(3, 5)$.
Find the equation of the line.

(3 marks)

You can write down the y-intercept of the line straight away.

Use a sketch to work out the gradient. Check the direction of the slope of the line to confirm whether the gradient is positive or negative.

Parallel and perpendicular

PARALLEL lines have the same gradient. These three lines all have a gradient of 1.

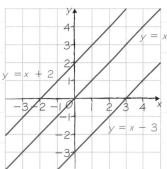

$y = x$

$y = x + 2$

$y = x - 3$

PERPENDICULAR means at right angles.

If a line has gradient m then any line perpendicular to it will have gradient $-\dfrac{1}{m}$.

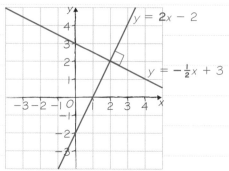

$y = 2x - 2$

$y = -\frac{1}{2}x + 3$

Worked example   grade A

A line L passes through the points $(-3, 6)$ and $(5, 4)$.

Another line, P, is perpendicular to L and passes through the point $(0, -7)$. Find the equation of line P.

Gradient of line L
$$= \frac{-2}{8} = \frac{-1}{4}$$

$(-3, 6)$ 8 -2 L $(5, 4)$

Gradient of line P
$$= \frac{-1}{\frac{-1}{4}} = 4$$

P passes through $(0, -7)$
Equation of P is: $y = 4x - 7$

1. Draw a sketch to find the gradient of line L.
2. The line slopes down so the gradient is negative.
3. Use $-\dfrac{1}{m}$ to calculate the gradient of P. If m is a fraction, you can just find its reciprocal and change the sign.
4. You know P passes through $(0, -7)$. Use $m = 4$ and $c = -7$ to write the equation of line P.

Check it!
If two lines are perpendicular the product of their gradients is -1: $-\frac{1}{4} \times 4 = -1$ ✓

Midpoints

A LINE SEGMENT is a short section of a straight line.

You can find the MIDPOINT of a line segment if you know the coordinates of the ends.

Coordinates of midpoint = (average of x-coordinates, average of y-coordinates)

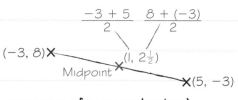

$\dfrac{-3 + 5}{2}$ $\dfrac{8 + (-3)}{2}$

$(-3, 8)$✕   ✕$(1, 2\frac{1}{2})$ Midpoint   ✕$(5, -3)$

Now try this

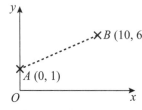

y
✕$B\ (10, 6)$
✕$A\ (0, 1)$
O x

A is the point $(0, 1)$.
B is the point $(10, 6)$.

edexcel

The equation of the straight line through A and B is $y = \frac{1}{2}x + 1$

(a) Write down the equation of another straight line that is parallel to AB. **(1 mark)**

(b) Write down the equation of another straight line that passes through the point $(0, 1)$. **(1 mark)**

(c) Find the equation of the line perpendicular to AB passing through B. **(2 marks)**

grade B

grade A

26

3-D coordinates

You can describe a point in three dimensions by using coordinates that have three numbers.

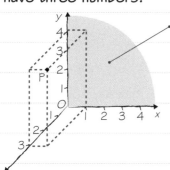

This is a normal coordinate grid with x- and y-axes. You can add a z-axis which is perpendicular to both the other axes.

You always give coordinates in the order (x, y, z).
The coordinates of point P are (1, 4, 3).
The arrow on each axis points in the positive direction.

Finding missing coordinates

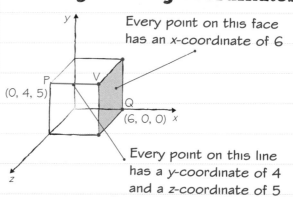

Every point on this face has an x-coordinate of 6

Every point on this line has a y-coordinate of 4 and a z-coordinate of 5

So the coordinates of point V are **(6, 4, 5)**.

Line segments

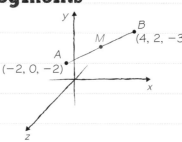

To find the coordinates of the midpoint of a line segment, you need to find the average of the coordinates of the endpoints.

The coordinates of M are
$$\left(\frac{-2 + 4}{2}, \frac{0 + 2}{2}, \frac{-2 + (-3)}{2}\right)$$
or (1, 1, $-2\frac{1}{2}$).

For a reminder about line segments in two dimensions look at page 26.

Worked example   grade **C**

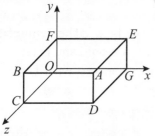

The diagram shows a cuboid drawn on a 3-D grid.
Vertex D has coordinates (6, 0, 5).
Vertex B has coordinates (0, 2, 5).
Write down the coordinates of vertex A.

(6, 2, 5)

Now try this

For part (b), write down the coordinates of C and S before finding the midpoint.

edexcel

A cuboid is shown on a 3-dimensional grid.

(a) Write down the letter of the point with coordinates (2, 0, 3).   **(1 mark)**

(b) Write down the coordinates of the midpoint of CS.   **(3 marks)**

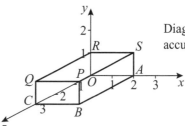

Diagram **NOT** accurately drawn

grade **B**

A*-A-B-C-D

Real-life graphs

Distance-time graphs

A DISTANCE–TIME graph shows how distance changes with time. This graph shows Jodi's run. The shape of the graph gives you information about the journey.

The gradient of the graph gives Jodi's speed.

$$\text{Gradient} = \frac{\text{distance up}}{\text{distance across}} = 1.9 \div \tfrac{1}{2} = 3.8$$

Jodi was travelling at 3.8 mph on this section of the run.

A horizontal line means no movement. Jodi rested here for 15 minutes.

Straight lines mean a constant speed.

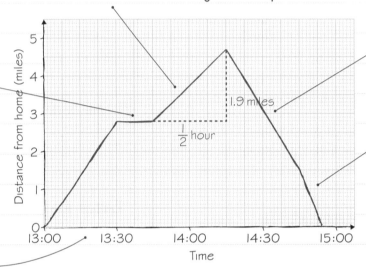

The horizontal scale might be marked in minutes or hours. Remember that there are 60 minutes in 1 hour.

Jodi sped up when she was nearly home. The graph is steeper here.

Graphs can be used to convert units or currencies, or to show how other quantities change with time. These garden ponds are filled with water at a constant rate. The graphs below show how the depth of water in each pond changes with time.

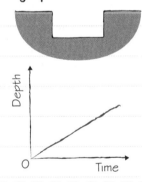

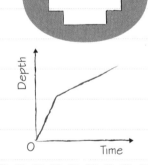

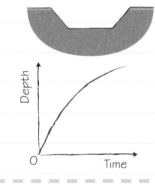

The narrower the section of pond, the faster the water depth will increase.

Now try this

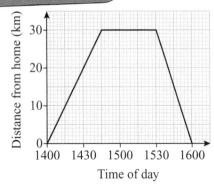

This distance–time graph shows Judy's journey to the airport and back.

(a) What is the distance from Judy's home to the airport? **(1 mark)**

(b) For how many minutes did Judy wait at the airport? **(1 mark)**

(c) Work out Judy's average speed on her journey home from the airport.
Give your answer in kilometres per hour. **(2 marks)**

grade **D**

grade **C**

-A*-
-A-
-B-
-C-
-D-

Formulae

A FORMULA is a mathematical rule.

You can write formulae using algebra.   FORMULAE is the plural of formula.

This label shows a formula for working out the cooking time of a chicken.

FREE-RANGE CHICKEN		
WEIGHT (KG) 1.8	**PRICE PER KG** £3.95	**COOKING INSTRUCTIONS** Cook at 170°C for 25 minutes per kg plus half an hour

You can write this formula using algebra as

$T = 25w + 30$, where T is the cooking time in minutes and w is the weight in kg.

In the description of each variable, you must give the units.

If T was the cooking time in hours then this formula would give you a very crispy chicken!

Worked example

grade **C**

This formula is used in physics to calculate distance.
$D = ut - 5t^2$
$u = 14$ and $t = -3$

Work out the value of D.

$D = (14)(-3) - 5(-3)^2$
$\quad = (14)(-3) - 5(9)$
$\quad = -42 - 45$
$\quad = -87$

Substitute the values for u and t into the formula.

If you use brackets then you're less likely to make a mistake. This is really important when there are negative numbers involved.

Remember **BIDMAS** for the correct order of operations. You need to do:

Indices → Multiplication → Subtraction

Don't try to do more than one operation on each line of working.

Worked example

grade **D**

Tom the plumber charges £35 for each hour he works at a job, plus £50. The amount Tom charges, in pounds, can be worked out using this rule.

> Multiply the number of hours he works by 35
> Add 50 to your answer

Tom works h hours at a job. He charges P pounds. Write down the formula for P in terms of h.

$P = 35h + 50$

Worked example

grade **D**

The cost of hiring a car is £100 plus £50 for each hire day.
Rita hires a car for d days.
Write down a formula for the total hire cost £C in terms of d.

Cost = £50 × number of days + £100
$C = 50d + 100$

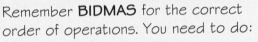

In these examples you need to turn a rule given in words into a formula. Sometimes it helps to write a formula in words first, then write it using algebra.

Now try this

grade **D**

(a) Work out the value of
$$\frac{p(q - 3)}{4}$$
when $p = 2$ and $q = -7$
(2 marks)

edexcel

(b) Pat plays a game with red cards and green cards. Red cards are worth 5 points each. Green cards are worth 3 points each. Pat has r red cards and g green cards. His total number of points is N. Write down a formula for N in terms of r and g.   **(3 marks)**

grade **C**

A*
A
B
C
D

Rearranging formulae

Most formulae have one letter on its own on one side of the formula. This letter is called the SUBJECT of the formula.

$$e = mc^2 \qquad e \text{ is the subject of the formula.}$$

CHANGING THE SUBJECT of a formula is like solving an equation. You have to do the same thing to both sides of the formula until you have the new letter on its own on one side.

$$e = mc^2 \qquad (\div m)$$
$$\frac{e}{m} = c^2 \qquad (\sqrt{})$$
$$\sqrt{\frac{e}{m}} = c \qquad c \text{ is now the subject of the formula.}$$

The inverse operation to is .
You need to square root EVERYTHING on both sides of the formula.

Harder formulae

If the letter you need APPEARS TWICE in the formula you need to FACTORISE.

| GROUP all the terms with that letter on one side of the formula and all the other terms on the other side. | → | FACTORISE so the letter only appears once. | → | DIVIDE by everything in the bracket to get the letter on its own. |

For a reminder about factorising have a look at page 22.

Worked example
grade C

$$N = \frac{3h + 20}{100}$$

Rearrange the formula to make h the subject.

$$N = \frac{3h + 20}{100} \qquad (\times 100)$$
$$100N = 3h + 20 \qquad (-20)$$
$$100N - 20 = 3h \qquad (\div 3)$$
$$\frac{100N - 20}{3} = h$$
$$h = \frac{100N - 20}{3}$$

It's a good idea to write your final answer as $h = ...$

Worked example
grade A

Make Q the subject of the formula $P = \dfrac{Q}{Q - 100}$.

$$P = \frac{Q}{Q - 100} \qquad [\times (Q - 100)]$$
$$P(Q - 100) = Q \qquad \text{(multiply out brackets)}$$
$$PQ - 100P = Q \qquad (+ 100P)$$
$$PQ = Q + 100P \, (- Q)$$
$$PQ - Q = 100P \qquad \text{(factorise)}$$
$$Q(P - 1) = 100P \qquad [\div (P - 1)]$$
$$Q = \frac{100P}{P - 1}$$

Your final answer should look like $Q = ...$
You need to factorise to get Q on its own.

Now try this

grade C

1. Make s the subject of the formula
   $v^2 = u^2 + 2as$ **(2 marks)**

2. Make x the subject of
   $5(x - 3) = y(4 - 3x)$ **(4 marks)**

grade A

edexcel

Inequalities

A*
A
B
C
D

An inequality tells you when one value is bigger or smaller than another value.
You can represent INEQUALITIES on a number line.

$x > -1$

−3 −2 −1 0 1 2 3 4

Use an OPEN circle for > and <

The open circle shows that −1 is NOT included.

$x \leqslant 3$

−3 −2 −1 0 1 2 3 4

Use a CLOSED circle for ⩾ and ⩽

The closed circle shows that 3 IS included.

Solving inequalities

You can solve an inequality in exactly the same way as you solve an equation.

$x - 3 \leqslant 12 \quad (+ 3)$

$\qquad x \leqslant 15$

The solution has the letter on its own on one side of the inequality and a number on the other side.

Golden rule

If you MULTIPLY or DIVIDE both sides of an inequality by a NEGATIVE number you have to REVERSE the INEQUALITY sign.

$6 - 5x > 10 \quad (- 6)$

$\quad -5x > 4 \quad (\div -5)$

$\qquad x < -\dfrac{4}{5}$

You have divided by a negative number so you have to reverse the inequality sign.

Worked example

grade **B**

Solve the <u>inequality</u> $\dfrac{2x}{3} < 10$

$\dfrac{2x}{3} < 10 \qquad (\times 3)$

$2x < 30 \qquad (\div 2)$

$\quad x < 15$

EXAM ALERT!

More than 60% of students got 0 marks on this question.

This is an **inequality** and not an equation.

You don't need to use an '=' sign in your answer.

This was a real exam question that caught students out – **be prepared!**

Results**Plus**

Integer solutions

You might need to write down all the integer solutions of an inequality.

INTEGERS are positive or negative whole numbers, including 0.

$-3 \leqslant x < 2$

−4 −3 −2 −1 0 1 2 3 4 5

This shows that x is between −3 and 2.
It can equal −3 but cannot equal 2.

The integer solutions to this inequality are −3, −2, −1, 0 and 1.

Now try this

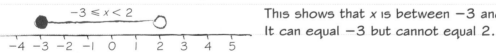

(a) Multiply each item by the LCM of 2 and 3.
(b) We want the smallest integer value of x that fits your answer to part (a).

edexcel

grade **C**

1. $-2 < x \leqslant 1$

x is an integer.

Write down all the possible values of x.

(2 marks)

2. (a) Solve the inequality

$\dfrac{3 + x}{2} > \dfrac{5 - 2x}{3}$

(b) x is an integer. Write down the smallest value of x that satisfies $\dfrac{3 + x}{2} > \dfrac{5 - 2x}{3}$

(4 marks)

grade **A**

-A*-
-A-
-B-
-C-
-D-

Inequalities on graphs

You can show the points that satisfy inequalities involving x and y on a graph.

For example, follow these steps to shade the region R that satisfies the inequalities:

$$x \geqslant 2 \qquad y > x \qquad x + y \geqslant 6$$

Always work on one inequality at a time.

1 $x \geqslant 2$
Draw the graph of $x = 2$ with a solid line. Use a small arrow to show which side of the line you want.

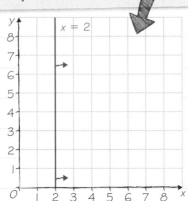

2 $y > x$
Draw the graph of $y = x$ with a dotted line.
Show which side of the line you want.

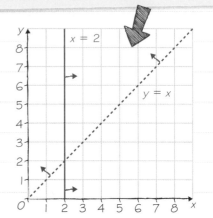

3 $x + y \geqslant 6$
Draw the graph of $x + y = 6$ with a solid line. Use a table of values.

x	0	3	6
y	6	3	0

Show which side of the line you want.
$x + y$ increases as you move away from the origin.
Shade in the region and label it **R**.

4   **Check it!**
Pick a point inside your shaded region. Check that the x- and y-values for that point satisfy **all** the inequalities.
At $(4, 5)$ $x = 4$ and $y = 5$.
$x \geqslant 2$ ✓
$y > x$ ✓
$x + y \geqslant 6$ ✓

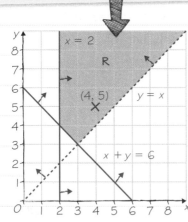

Graphical inequalities checklist

$<$ and $>$ are shown by DOTTED lines. ✓

$\leqslant$ and $\geqslant$ are shown by SOLID lines. ✓

Points on a solid line ARE included in the region. ✓

Points on a dotted line AREN'T included in the region. ✓

Now try this

edexcel

On a grid with $-5 \leqslant x \leqslant 5$ and $-2 \leqslant y \leqslant 5$, mark with a cross (×) each of the six points which satisfy all of these three inequalities where x and y are both integers.

$x \geqslant -2 \qquad y \geqslant 1 \qquad x + y < 2$

(4 marks)

grade
A

Quadratic and cubic graphs

A*
A
B
C
D

An equation with an x^2 term is a QUADRATIC.

An equation with an x^3 term is a CUBIC.

You can use a table of values to draw graphs of quadratic and cubic equations.

$y = x^2 - 3x + 1$

x	−1	0	1	2	3	4
y	5	1	−1	−1	1	5

All quadratic graphs have a line of symmetry which passes through the minimum or maximum point.

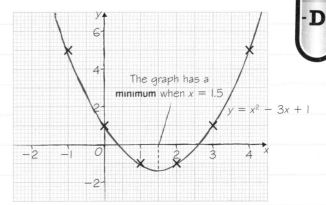

The graph has a minimum when $x = 1.5$

$y = x^2 - 3x + 1$

The graph of $y = x^2 - 3x + 1$ is symmetrical about the line $x = 1.5$

Worked example

grade B

This is a graph of $y = 2x^2 + 5x$

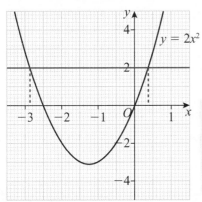

$y = 2x^2 + 5x$

Everything in red is part of the answer.

By drawing a suitable straight line on the graph, solve the equation $2x^2 + 5x - 2 = 0$

Give your answers correct to 1 decimal place.

$2x^2 + 5x - 2 = 0 \quad (+2)$
$2x^2 + 5x = 2$
$x = 0.4, x = -2.9$

You can solve the **quadratic equation** $2x^2 + 5x - 2 = 0$ by finding where the graph $y = 2x^2 + 5x$ crosses the straight line $y = 2$.

Draw the line $y = 2$ on the graph.

The solutions are the x-values at the points of intersection.

You can revise solving a similar pair of equations algebraically on page 42.

Cubic graphs

You need to know the general shape of a cubic graph if you're going for a grade A.

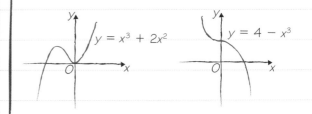

$y = x^3 + 2x^2$ $y = 4 - x^3$

Now try this

grade B   edexcel

(a) Complete the table of values for $y = 2x^2 - 4x$

x	−2	−1	0	1	2	3
y	16		0			6

(2 marks)

(b) On a grid with $-2 \leqslant x \leqslant 3$ and $-5 \leqslant y \leqslant 20$, draw the graph of $y = 2x^2 - 4x$ for values of x from −2 to 3   **(2 marks)**

(c) Write down the values of x for which $2x^2 - 4x - 3 = 0$ **(2 marks)**

A*
A
B
C
D

Graphs of $\frac{k}{x}$ and a^x

1 Graphs of the form $y = \frac{k}{x}$ are called RECIPROCAL GRAPHS.

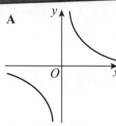
$y = \frac{1}{x}$

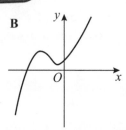
$y = -\frac{2}{x}$

For quadratic and cubic graphs see page 33.

2 Graphs of the form $y = a^x$ or $y = a^{-x}$ are called EXPONENTIAL GRAPHS.

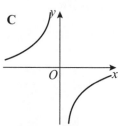
$y = 2^x$
$y = \left(\frac{1}{2}\right)^x$

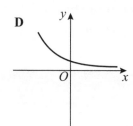
$y = 3^{-x}$
$y = 2^{-x}$

Exponentials sometimes appear in equations representing growth or decay in nature.

Worked example

grade
A

A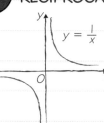

B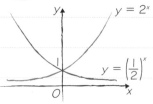

C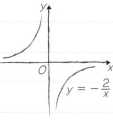

D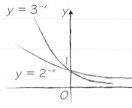

Write down the letter of the graph which could have the equation

(a) $y = 0.5^{-x}$ ....D....   (b) $y = \frac{2}{x}$ ...A...   (c) $y = x^3 + 3x^2 + 2x + 1$ ....B...

Finding missing values

Some questions ask you to find the missing value in an exponential equation.

This graph has equation $y = ka^x$. You can substitute the values of x and y that you are given to get two equations.

$$7 = ka^1 \qquad (1)$$
$$175 = ka^3 \qquad (2)$$
$$ka^3 \div ka^1 = 175 \div 7 \qquad (2) \div (1)$$
$$a^2 = 25$$
$$a = 5$$

Substitute $a = 5$ into (1):
$$7 = 5k$$
$$k = \frac{7}{5}$$

Check it!
Substitute in (2):
$$ka^3 = \frac{7}{5}(5)^3$$
$$= 7 \times 25$$
$$= 175 \checkmark$$

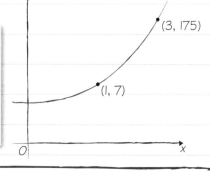
(3, 175)
(1, 7)

Now try this

edexcel

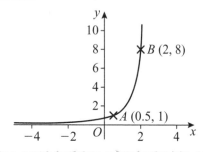
B (2, 8)
A (0.5, 1)

The diagram shows a sketch of the graph $y = ab^x$.
The curve passes through the points $A(0.5, 1)$ and $B(2, 8)$.
The point $C(-0.5, k)$ lies on the curve.
Find the value of k.   **(4 marks)**

grade
A*

Trial and improvement

Some equations can't be solved exactly. You need to use trial and improvement to find an approximate solution. You will be told when to use trial and improvement in your exam. Look at the worked example below which shows you how to do it in two steps.

Worked example

1 The equation $x^3 - 5x = 60$ has a solution between 4 and 5
Use a trial and improvement method to find this solution.
Give your answer correct to 1 decimal place.

x	$x^3 - 5x$	Too big or too small
4.5	68.625	Too big
4.2	53.088	Too small

Draw a table to record your working. You know there is a solution between 4 and 5.

$x = 4.5$ is a good first value to try.

Use your calculator to work out $4.5^3 - 5 \times 4.5$ and compare your answer with 60.

$x = 4.5$ is too big.

Try $x = 4.2$

2 The equation $x^3 - 5x = 60$ has a solution between 4 and 5
Use a trial and improvement method to find this solution.
Give your answer correct to 1 decimal place.

x	$x^3 - 5x$	Too big or too small
4.5	68.625	Too big
4.2	53.088	Too small
4.3	58.007	Too small
4.4	63.184	Too big
4.35	60.56...	Too big

$x = 4.3$ (to 1 d.p.)

Keep trying different values.

Make sure you write down the result of every trial.

You know the answer is between 4.3 and 4.4. But you don't know which value is closer.

Try 4.35. This will tell you whether the answer is closer to 4.3 or 4.4

4.35 is too big so the answer is between 4.3 and 4.35

Write down the answer correct to 1 decimal place.

Now try this

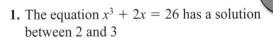

edexcel

1. The equation $x^3 + 2x = 26$ has a solution between 2 and 3

Use a trial and improvement method to find this solution.

Give your answer correct to 1 decimal place.

You must show **all** your working.

(4 marks)

2. The equation $x^3 + 4x^2 = 100$ has a solution between 3 and 4

Use a trial and improvement method to find this solution.

Give your answer correct to 1 decimal place.

You must show **all** your working.

(4 marks)

-A*-
-A-
-B-
-C-
-D-

Simultaneous equations 1

Simultaneous equations have two unknowns. You need to find the values for the two unknowns which make BOTH equations true.

Algebraic solution

1. Number each equation.

2. If necessary, multiply the equations so that the coefficients of one unknown are the same.

3. Add or subtract the equations to ELIMINATE that unknown.

4. Once one unknown is found use substitution to find the other.

5. Check the answer by substituting both values into the original equations.

$$3x + y = 20 \quad (1)$$
$$x + 4y = 14 \quad (2)$$

$$12x + 4y = 80 \quad (1) \times 4$$
$$- (x + 4y = 14) \quad - (2)$$
$$\overline{}$$
$$11x = 66$$
$$x = 6$$

Substitute $x = 6$ into (1):
$$3(6) + y = 20$$
$$18 + y = 20$$
$$y = 2$$

Solution is $x = 6$, $y = 2$
Check: $x + 4y = 6 + 4(2) = 14$ ✓

Worked example

grade A

Solve the simultaneous equations
$$6x + 2y = -3 \quad (1)$$
$$4x - 3y = 11 \quad (2)$$

$$18x + 6y = -9 \quad (1) \times 3$$
$$+ 8x - 6y = 22 \quad (2) \times 2$$
$$\overline{}$$
$$26x = 13$$
$$x = \tfrac{1}{2}$$

Substitute $x = \tfrac{1}{2}$ into (1):
$$6\left(\tfrac{1}{2}\right) + 2y = -3$$
$$3 + 2y = -3$$
$$2y = -6$$
$$y = -3$$

EXAM ALERT!

Only 1 in 6 students got full marks in this question.

When deciding which unknown to eliminate, if possible choose the unknown where the signs are different. You can then eliminate the unknown by adding the equations.

Multiply both equations by a whole number to make the coefficients the same.

Check it!
Always use the equation you didn't substitute into to check.
$$4x - 3y = 4\left(\tfrac{1}{2}\right) - 3(-3) = 2 + 9 = 11 ✓$$

This was a real exam question that caught students out – **be prepared!**   Results Plus

Graphical solution

You can solve these simultaneous equations by drawing a graph.
$$x - y = 1 \qquad x + 2y = 4$$
The coordinates of the point of intersection give the solution to the simultaneous equations.

The solution is $x = 2$, $y = 1$.

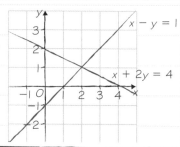

Now try this

grade B

1. Solve the simultaneous equations
$$2y - 3x = 6$$
$$2x + 2y = 1 \qquad \textbf{(4 marks)}$$

2. Solve the simultaneous equations
$$2x + 3y = -3$$
$$3x - 2y = 28 \qquad \textbf{(4 marks)}$$

grade A

edexcel

A*
A
B
C
D

Quadratic equations

Quadratic equations can be written in the form $ax^2 + bx + c = 0$ where a, b and c are numbers.

If a quadratic equation is given in a different form you should rearrange it before solving it.

If you can factorise the left-hand side of a quadratic equation, you can solve it without a calculator.

For a reminder about factorising quadratic expressions have a look at page 22.

Two to watch

1. When $c = 0$:
$$x^2 - 10x = 0$$
$$x(x - 10) = 0$$
Solutions are $x = 0$ and $x = 10$.

2. When $b = 0$ (difference of two squares):
$$9x^2 - 4 = 0$$
$$(3x + 2)(3x - 2) = 0$$
Solutions are $x = \frac{2}{3}$ and $x = -\frac{2}{3}$.

Worked example   grade B

Solve $x^2 + 8x - 9 = 0$

$$(x + 9)(x - 1) = 0$$

$$x + 9 = 0 \qquad x - 1 = 0$$
$$x = -9 \qquad x = 1$$

Quadratic equations questions can go up to A*. You might have to form the equation before solving the problem.

EXAM ALERT!

Just over half of students got this question wrong.

To solve by factorising, look for two numbers which add up to 8 and multiply to make -9. The numbers are 9 and -1.

Set each factor to 0 then solve to find the solutions.

Check it!
$(1)^2 + 8(1) - 9 = 1 + 8 - 9 = 0$ ✓
$(-9)^2 + 8(-9) - 9 = 81 - 72 - 9 = 0$ ✓

This was a real exam question that caught students out – **be prepared!**   ResultsPlus

Quadratic equations sometimes appear when you are solving problems.

What is the value of x in this diagram?

←— x cm —→
($x + 3$) cm
x cm   Area = 20 cm²
←——— $2x$ cm ———→

$$x^2 + x(x + 3) = 20 \quad\longrightarrow\quad \text{Write an expression for the area in terms of } x \text{ and set it equal to 20}$$

$$x^2 + x^2 + 3x = 20$$

$$2x^2 + 3x - 20 = 0 \quad\longrightarrow\quad \text{This is a quadratic equation. Rearrange it into the form } ax^2 + bx + c = 0$$

$$(2x - 5)(x + 4) = 0$$

$$2x - 5 = 0 \qquad x + 4 = 0$$
$$x = 2\tfrac{1}{2} \qquad x = -4$$

$$\text{So } x = 2\tfrac{1}{2}$$

Only one of these solutions makes sense. Lengths can't be negative numbers.

Now try this

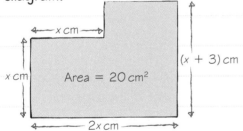
$x + 2$
$x - 5$
$x + 6$

Diagram **NOT** accurately drawn

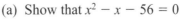 edexcel

grade A

The diagram shows a trapezium. All measurements are given in centimetres. The area of the trapezium is 36 cm².

(a) Show that $x^2 - x - 56 = 0$ **(2 marks)**

(b) (i) Solve the equation $x^2 - x - 56 = 0$ **(3 marks)**
     (ii) Hence find the length of the shortest side of the trapezium.   **(1 mark)**

A* A B C D

Completing the square

If a quadratic expression is written in the form $(x + p)^2 + q$, it is in COMPLETED SQUARE FORM.

Completing the square

You can use this formula to complete the square:

$$x^2 + 2bx + c = (x + b)^2 - b^2 + c$$

Minimum values

Quadratic graphs are curves. This quadratic graph has a minimum point. You can find the coordinates of the minimum point by completing the square.

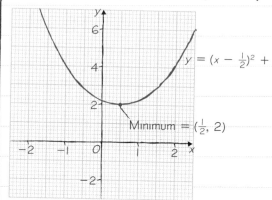

$y = (x - \frac{1}{2})^2 + 2$

Minimum $= (\frac{1}{2}, 2)$

The minimum point of the graph $y = (x + p)^2 + q$ is at $(-p, q)$.

Learn this if you're going for an A*.

 Worked example grade A

(a) Write $x^2 + 6x + 20$ in the form $(x + p)^2 + q$

$$\begin{aligned} x^2 + 6x + 20 &= (x + 3)^2 - 3^2 + 20 \\ &= (x + 3)^2 - 9 + 20 \\ &= (x + 3)^2 + 11 \end{aligned}$$

(b) What is the minimum value of $x^2 + 6x + 20$?

Minimum value = 11

(a) Look at the formula for completing the square.

$2b = 6$ so $b = 3$ and $c = 20$

Substitute these values into the formula.

(b) $(x + 3)^2$ is a square, so it can't be negative. The minimum value for $(x + 3)^2$ is 0, so the minimum value for $(x + 3)^2 + 11$ is 11.

Check it!

Multiply out the brackets:

$$\begin{aligned} (x + 3)^2 + 11 &= x^2 + 3x + 3x + 9 + 11 \\ &= x^2 + 6x + 20 \checkmark \end{aligned}$$

Solving the equation

Once you have completed the square, the unknown only appears once in the equation. You can solve the equation using inverse operations.

$$\begin{aligned} x^2 - 8x + 1 &= 0 \\ (x - 4)^2 - 16 + 1 &= 0 \\ (x - 4)^2 - 15 &= 0 \qquad (+ 15) \\ (x - 4)^2 &= 15 \qquad (\sqrt{}) \\ x - 4 &= \pm \sqrt{15} \qquad (+ 4) \\ x &= 4 \pm \sqrt{15} \end{aligned}$$

When you square root both sides, you need to add a $\pm$ sign in front of the square root.

Completing the square is a useful method for solving equations when you need to give your answer as a SURD.

 Now try this

Divide each term by 2 before you start.

edexcel

1. Show that $x^2 - 4x + 15$ can be written as $(x + p)^2 + q$ for all values of x.
State the values of p and q.   **(3 marks)**

grade A

2. Solve $2x^2 + 8x - 3 = 0$
Give your answers in the form $p \pm \sqrt{\dfrac{q}{r}}$ where p, q and r are integers.   **(4 marks)**

grade A

The quadratic formula

This is how the quadratic formula will appear on the formula sheet in your exam.

> **The Quadratic Equation**
>
> The solutions of $ax^2 + bx + c = 0$
>
> where $a \neq 0$, are given by
>
> $$x = \frac{-b \pm \sqrt{(b^2 - 4ac)}}{2a}$$

If you're going for an A or A*, you may need to use this in a problem-solving question.

Safe substituting

Equation is in the form $ax^2 + bx + c = 0$. ✓

Write down your values of a, b and c before you substitute. ✓

Use brackets when you are substituting negative numbers. ✓

Show what you have substituted in the formula. ✓

Simplify what is under the square root and write this down. ✓

The $\pm$ symbol means you need to do two calculations. ✓

Worked example

grade A

Solve $5x^2 + x + 11 = 14$

Give your solutions correct to 3 significant figures.

$5x^2 + x - 3 = 0$

$a = 5, b = 1, c = -3$

$$x = \frac{-1 \pm \sqrt{1^2 - 4 \times 5 \times (-3)}}{2 \times 5}$$

$$= \frac{-1 + \sqrt{61}}{10} \text{ or } \frac{-1 - \sqrt{61}}{10}$$

$$= 0.681024... \text{ or } -0.881024...$$

$$= 0.681 \text{ or } -0.881 \text{ (to 3 s.f.)}$$

You are asked to find 'solutions'. This tells you that you are solving a quadratic equation.

You must give your answer 'correct to 3 significant figures'. This tells you that you need to use the quadratic formula. Turn to the formula sheet.

Write down at least five figures after the decimal point on the calculator display before giving your final answer. You might need to use the ⓢ⇔ⓓ button on your calculator to get your answer as a decimal.

How many solutions?

A quadratic equation can have two solutions, one solution or no solutions. You can use $b^2 - 4ac$ (the part under the square root) to work out how many solutions a quadratic equation has.

If $b^2 - 4ac$ is negative, there are no solutions.

If $b^2 - 4ac = 0$, there is only one solution.

If $b^2 - 4ac > 0$, there are two different solutions.

You can't calculate the square root of a negative number.

± 0 appears in the formula, so you get the same answer whether you use $+$ or $-$.

Now try this

edexcel

1. Solve $3x^2 + 7x - 13 = 0$
Give your solutions correct to 2 decimal places. **(3 marks)**

grade A

2. Solve $x^2 + x + 11 = 14$
Give your solutions correct to 3 significant figures. **(3 marks)**

grade A

Quadratics and fractions

You need to remove fractions before you can solve an equation.
For a reminder about solving linear equations with fractions have a look at page 24.

To remove fractions from an equation multiply everything by the lowest common multiple of the denominators.

$$\frac{x}{2x-3} + \frac{4}{x+1} = 1$$

> (2x − 3) and (x + 1) don't have any common factors. Multiply everything by (2x − 3)(x + 1).

$$\frac{x\cancel{(2x-3)}(x+1)}{\cancel{2x-3}} + \frac{4(2x-3)\cancel{(x+1)}}{\cancel{x+1}} = (2x-3)(x+1)$$

> Don't expand brackets until you have simplified the fractions.

$$x(x+1) + 4(2x-3) = (2x-3)(x+1)$$
$$x^2 + x + 8x - 12 = 2x^2 - 3x + 2x - 3$$
$$0 = x^2 - 10x + 9$$
$$= (x-9)(x-1)$$

> Multiply out brackets and collect like terms.

Solutions are $x = 9$ and $x = 1$.

Worked example

grade A*

Solve the equation $\dfrac{5}{x+2} = \dfrac{4-3x}{x-1}$

Give your solutions correct to 2 decimal places.

$$\left(\frac{5}{x+2}\right)(x+2)(x-1) = \left(\frac{4-3x}{x-1}\right)(x+2)(x-1)$$

$$\frac{5\cancel{(x+2)}(x-1)}{\cancel{x+2}} = \frac{(4-3x)(x+2)\cancel{(x-1)}}{\cancel{x-1}}$$

$$5(x-1) = (4-3x)(x+2)$$
$$5x - 5 = 4x - 3x^2 + 8 - 6x$$
$$3x^2 + 7x - 13 = 0$$

$$a = 3, b = 7, c = -13$$

$$x = \frac{-7 \pm \sqrt{7^2 - 4 \times 3 \times (-13)}}{2 \times 3}$$

$$= \frac{-7 + \sqrt{205}}{6} \text{ or } \frac{-7 - \sqrt{205}}{6}$$

$$= 1.21963\ldots \text{ or } -3.55297\ldots$$

$$= 1.22 \qquad \text{or } -3.55 \text{ (to 2 d.p.)}$$

Quadratic equations checklist

Remove any fractions by multiplying everything by the lowest common multiple of the denominators. ✓

Multiply out any brackets and collect like terms. ✓

Rewrite in the form $ax^2 + bx + c = 0$. ✓

Factorise if possible. ✓

Use the quadratic formula when asked to round the answer. ✓

> The question asks for your solutions correct to 2 decimal places so you will need to use the quadratic formula.

Now try this

> First factorise $x^2 - 9$ (difference of two squares).
> Your lowest common multiple will now only have 2 brackets.

grade A*

1. Solve the equation $\dfrac{3}{x+1} - \dfrac{2}{x+5} = 1$

Give your answers correct to 2 decimal places.

(5 marks)

2. Solve the equation

$$\frac{3}{x+3} - \frac{4}{x-3} = \frac{5x}{x^2-9}$$

(4 marks)

grade A*

edexcel

Equation of a circle

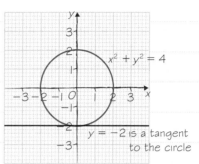

A circle of radius r centred at the origin has equation $x^2 + y^2 = r^2$.

This circle is the locus of points which are a distance r from the origin.

A line which just touches the circle once is called a tangent.

Always draw circles with a pair of compasses.

$x^2 + y^2 = 4$

$y = -2$ is a tangent to the circle

Worked example

grade A*

The diagram shows a circle of radius 4 with its centre at the origin.

By drawing a suitable line on the diagram estimate the solutions of the simultaneous equations:
$x^2 + y^2 = 16$
$y = -\frac{1}{2}x + 2$

Solutions: $x = 4$, $y = 0$ and
$x = -2.4$, $y = 3.2$

Everything in red is part of the answer.

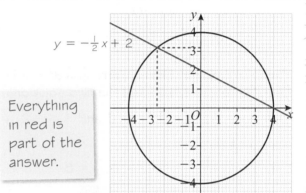

$y = -\frac{1}{2}x + 2$

$x^2 + y^2 = 16$ represents a circle of radius 4 with its centre at (0, 0). This is the circle which has been drawn on the diagram.

Draw the line $y = -\frac{1}{2}x + 2$ on the same diagram. This is a straight line with gradient $-\frac{1}{2}$ and y-intercept (0, 2). The solutions are the coordinates of the points of intersection.

Drawing circles

You can draw your own graphs to estimate the solutions of these simultaneous equations.

$x^2 + y^2 = 100$ ⟶ $x^2 + y^2 = 100$ is a circle of radius 10 with its centre at (0, 0).

$2y = 3x - 4$

This is a straight line so use a table of values.

x	−4	0	4
y	−8	−2	4

Estimated solutions:
$x = 6.4$, $y = 7.8$ and $x = -4.6$, $y = -8.8$.

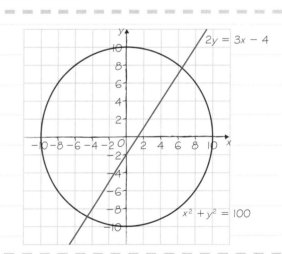

$2y = 3x - 4$

$x^2 + y^2 = 100$

Now try this

grade A*   edexcel

A circle has equation $x^2 + y^2 = 9$

A straight line has equation $y = x + 5$

Show graphically that there are no solutions to the simultaneous equations $x^2 + y^2 = 9$ and $y = x + 5$ **(3 marks)**

Draw the given circle and the given straight line on a graph. Write down what you notice.

-A*-
-A-
-B-
-C-
-D-

Simultaneous equations 2

If there is an x^2 or y^2 term in a pair of simultaneous equations, you need to solve them using SUBSTITUTION.

$$y = x^2 - 2x - 7 \qquad (1)$$
$$x - y = -3 \qquad (2)$$

Rearrange the linear equation to make one letter the subject.

$$y = x + 3 \qquad (3)$$

Substitute (3) into (1):

$$x + 3 = x^2 - 2x - 7$$
$$0 = x^2 - 3x - 10$$
$$0 = (x - 5)(x + 2)$$

$$x = 5 \text{ or } x = -2$$

Each solution for x has a corresponding value of y. Substitute into (3) to find the two solutions.

Solutions are $x = 5$, $y = 8$ and $x = -2$, $y = 1$.

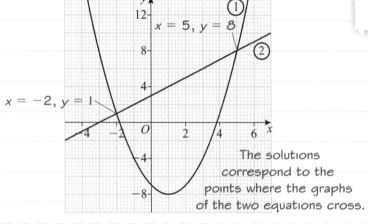

The solutions correspond to the points where the graphs of the two equations cross.

grade
A*

Worked example

Solve the simultaneous equations
$$x - 2y = 1 \qquad (1)$$
$$x^2 + y^2 = 13 \qquad (2)$$

$$x = 1 + 2y \qquad (3)$$

Substitute (3) into (2):

$$(1 + 2y)^2 + y^2 = 13$$
$$1 + 4y + 4y^2 + y^2 = 13$$
$$5y^2 + 4y - 12 = 0$$
$$(5y - 6)(y + 2) = 0$$

$$y = \frac{6}{5} \text{ or } y = -2$$

$$x = 1 + 2\left(\frac{6}{5}\right) \qquad\qquad x = 1 + 2(-2)$$
$$= \frac{17}{5} \qquad\qquad\qquad = -3$$

Solutions: $x = \frac{17}{5}$, $y = \frac{6}{5}$ and
$$x = -3, \ y = -2.$$

You can substitute for x or y. It is easier to substitute for x because there will be no fractions.

Use brackets to make sure that the whole expression is squared.

Rearrange the quadratic equation for y into the form $ay^2 + by + c = 0$.

Factorise the left-hand side to find two solutions for y.

How many solutions?

When one equation is linear and the other is quadratic there can be one solution, two solutions or no solutions.

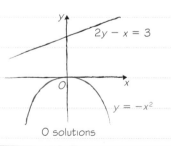

| 1 solution | 2 solutions | 0 solutions |

Now try this

grade
A*

edexcel

1. Solve the simultaneous equations
$$x^2 + y^2 = 29$$
$$y - x = 3 \qquad \textbf{(7 marks)}$$

2. Solve the simultaneous equations
$$y + 1 = x^2$$
$$x = y - 1 \qquad \textbf{(6 marks)}$$

grade
A*

Direct proportion

Direct proportion was introduced on page 12. This page looks at it in more detail.

The relationship between the cost of petrol and the number of litres you buy is an example of direct proportion.

If you buy *L* litres for £*C* then you can write

- a statement of proportionality: $C \propto L$
  $\propto$ means 'is proportional to'.

- a formula for direct proportion: $C = kL$
  k is called the CONSTANT OF PROPORTIONALITY.

Direct proportion graphs

Straight line. ✓
Passes through the origin. ✓
k is the gradient of the line. ✓

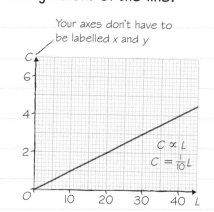

Your axes don't have to be labelled *x* and *y*

$C \propto L$
$C = \frac{1}{10}L$

Worked example grade **A**

Winnie drops a stone down a well. The speed of the stone, *v* m/s, is directly proportional to the time, *t* seconds, since she dropped it.

After 0.5 seconds the stone is travelling at 4.9 m/s.

(a) Find a formula for *v* in terms of *t*.

$v = kt$
$4.9 = k(0.5)$ $(\div 0.5)$
$k = 9.8\, \text{m/s}^2$
$v = 9.8t$

(b) Calculate the speed of the stone after 1.2 seconds.

$v = 9.8(1.2) = 11.76\, \text{m/s}$

1. Write down the formula using *k* for the constant of proportionality.
2. Substitute the values of *v* and *t* you are given.
3. Solve the equation to find the value of *k*.
4. Write down the formula putting in the value of *k*.
5. Once you have written your formula, you can use it to find the value of one variable if you know the value of the other.

Checking for proportionality

You can use a graph to check whether two quantities are directly proportional.

P	5	10	15
Q	1.2	1.5	1.8

The graph doesn't go through the origin so P and Q are not directly proportional.

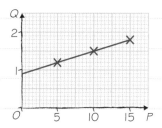

Now try this   edexcel   grade **A**

A ball falls vertically after being dropped. The ball falls a distance *d* metres in a time of *t* seconds.

d is directly proportional to the square of *t*.

The ball falls 20 metres in a time of 2 seconds.

(a) Find a formula for *d* in terms of *t*. **(3 marks)**
(b) Calculate the distance the ball falls in 3 seconds. **(2 marks)**
(c) Calculate the time the ball takes to fall 605 m. **(2 marks)**

Proportionality formulae

You can answer some tricky proportionality questions quickly by remembering the proportionality FORMULAE and the shapes of the proportionality GRAPHS.

Proportionality in words	Using $\propto$	Formula
y is directly proportional to x	$y \propto x$	$y = kx$
y is directly proportional to the square of x	$y \propto x^2$	$y = kx^2$
y is directly proportional to the cube of x	$y \propto x^3$	$y = kx^3$
y is directly proportional to the square root of x	$y \propto \sqrt{x}$	$y = k\sqrt{x}$
y is inversely proportional to x	$y \propto \dfrac{1}{x}$	$y = \dfrac{k}{x}$
y is inversely proportional to the square of x	$y \propto \dfrac{1}{x^2}$	$y = \dfrac{k}{x^2}$

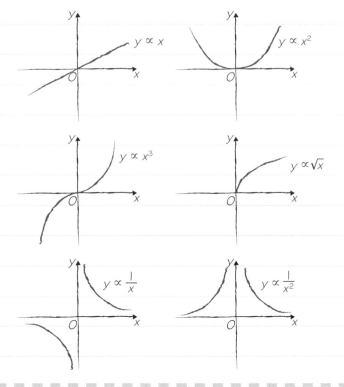

Worked example

grade A

q is <u>inversely proportional</u> to the square of t.

When $t = 4$, $q = 8.5$

Calculate the value of q when $t = 5$

$q \propto \dfrac{1}{t^2}$

$q = \dfrac{k}{t^2}$

$8.5 = \dfrac{k}{4^2}$

$k = 8.5 \times 4^2 = 136$

$q = \dfrac{136}{t^2}$

When $t = 5$: $q = \dfrac{136}{5^2} = 5.44$

Always...

1. Write down the statement of proportionality and then the formula.
2. Substitute the values you are given.
3. Solve the equation to find k.
4. Write down the formula using the value of k.
5. Use your formula to find any unknown values.

⬅ Don't round your answer unless the question tells you to.

Now try this

grade A

edexcel

q is inversely proportional to the square root of t.

When $t = 9$, $q = 4$

Calculate the value of q when $t = 100$ **(4 marks)**

⬅ The key words are 'inversely' and 'square root'.

Transformations 1

You can change the equation of a graph to translate it, stretch it or reflect it.
In the exam you might have to use functions to describe these transformations.

Function	$y = f(x) + a$	$y = f(x + a)$	$y = af(x)$
Transformation of graph	Translation $\begin{pmatrix} 0 \\ a \end{pmatrix}$	Translation $\begin{pmatrix} -a \\ 0 \end{pmatrix}$	Stretch in the vertical direction, scale factor a
Useful to know	$f(x) + a \rightarrow$ move UP a units $\quad$ $f(x) - a \rightarrow$ move DOWN a units	$f(x + a) \rightarrow$ move LEFT a units $\quad$ $f(x - a) \rightarrow$ move RIGHT a units	x-values stay the same
Example	$y = f(x) + 3$ $y = f(x)$	$y = f(x)$ $y = f(x + 5)$	$y = 3f(x)$ $y = f(x)$

Function	$y = f(ax)$	$y = -f(x)$	$y = f(-x)$
Transformation of graph	Stretch in the horizontal direction, scale factor $\frac{1}{a}$	Reflection in the x-axis	Reflection in the y-axis
Useful to know	y-values stay the same	'$-$' outside the bracket	'$-$' inside the bracket
Example	$y = f(2x)$ $y = f(x)$	$y = f(x)$ $y = -f(x)$	$y = f(-x)$ $\quad$ $y = f(x)$

Worked example

grade **A***

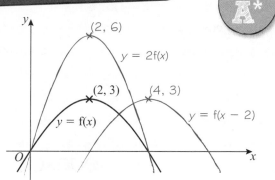

The curve $y = f(x)$ has a vertex at $(2, 3)$.
Write down the coordinates of the vertex of the curve with equation
(a) $y = f(x - 2)$ $(4, 3)$
(b) $y = 2f(x)$ $(2, 6)$

$y = f(x - 2)$ is a translation 2 units right along the x-axis.
$y = 2f(x)$ is a stretch in the vertical direction, scale factor 2.

Now try this

edexcel

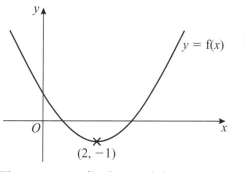

grade **A***

The curve $y = f(x)$ has a minimum point at $(2, -1)$.

(a) Write down the coordinates of the minimum point of the curve with equation
   (i) $y = f(x + 2)$
   (ii) $y = 3f(x)$
   (iii) $y = f(2x)$ **(3 marks)**

The curve $y = f(x)$ is reflected in the y-axis.

(b) Find the equation of the curve following this transformation.   **(1 mark)**

A*
A
B
C
D

Transformations 2

You need to be able to convert between FUNCTION NOTATION and equations of graphs. This table shows some transformations that may come up in your exam.

Original function	$y = 2x + 3$	$y = \sin x°$	$y = x^2 - 2x + 1$	$y = x^2$
Transformation	$f(x) \rightarrow f(x) + 2$	$f(x) \rightarrow f(x - 30)$	$f(x) \rightarrow 2f(x)$	$f(x) \rightarrow f(3x)$
Which means...	movement UP by 2 units	movement RIGHT by 30°	stretch in vertical direction, scale factor 2	stretch in horizontal direction, scale factor $\frac{1}{3}$
New function	$y = 2x + 5$	$y = \sin(x - 30)°$	$y = 2x^2 - 4x + 2$	$y = 9x^2$

Graphs of sine and cosine

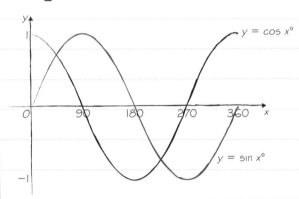

The graph of $y = \cos x°$ is identical to the graph of $y = \sin x°$ except that it has been moved to the left by 90°.

Write down the transformations using function notation.
(a) Stretch in the vertical direction with scale factor $\frac{1}{2}$.
(b) Stretch in the horizontal direction with scale factor 2.

Worked example

grade A*

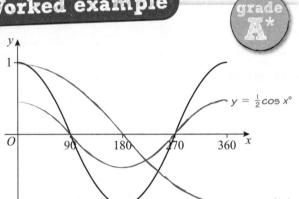

The diagram shows a sketch of the curve $y = \cos x°$ for $0 \leqslant x \leqslant 360$

On the same diagram sketch the curve with equation

(a) $y = \frac{1}{2} \cos x°$ $y = \frac{1}{2}f(x)$

(b) $y = \cos \left(\frac{1}{2} x\right)°$ $y = f(\frac{1}{2}x)$

Now try this

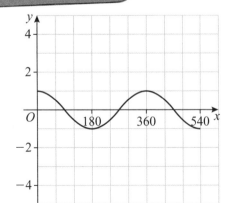

edexcel

The grid shows the graph of $y = \cos x°$ for values of x from 0 to 540

On the grid, sketch the graph of $y = 3\cos(2x)°$ for values of x from 0 to 540   **(2 marks)**

grade A*

Algebraic fractions

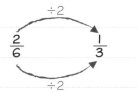

-A*-
-A-
-B-
-C-
-D-

Simplifying an algebraic fraction is just like simplifying a normal fraction.

$$\frac{2}{6} \xrightarrow{\div 2} \frac{1}{3}$$

$$\frac{x+1}{2x(x+1)} \xrightarrow{\div(x+1)} \frac{1}{2x}$$

You can divide the top and bottom of the fraction by a number, a term, or a whole expression.

Remember you have to divide by exactly the same thing on the top AND the bottom.

Golden rule

If the top or the bottom of the fraction has MORE THAN one term, you will need to factorise before simplifying.

$$\frac{p^2 + 3p}{4p} = \frac{p(p+3)}{4p} = \frac{p+3}{4}$$

Two terms on top so factorise the top, then divide the top and bottom by p.

There is more on dividing algebraic terms on page 19.

For a reminder on how to factorise an expression see page 22.

Operations on algebraic fractions

1 To ADD or SUBTRACT algebraic fractions with different denominators:

1. Find a common denominator.   2. Add or subtract the numerators.

3. Don't change the denominator.   4. Simplify if possible.

$$\frac{1}{x+4} + \frac{2}{x-4} = \frac{x-4}{(x+4)(x-4)} + \frac{2(x+4)}{(x+4)(x-4)}$$
$$= \frac{x-4+2x+8}{(x+4)(x-4)} = \frac{3x+4}{(x+4)(x-4)}$$

The smallest common denominator isn't always the product of the two denominators.

You can use a common denominator of $4x$ to simplify this expression: $\dfrac{x+1}{2x} + \dfrac{3-2x}{4x}$

2 To MULTIPLY fractions:

1. Multiply the numerators AND multiply the denominators.

2. Simplify if possible.

$$\frac{x}{2} \times \frac{4}{x-1} = \frac{{}^{2}4x}{{}_{1}2(x-1)} = \frac{2x}{x-1}$$

Don't expand brackets if you don't have to. It's much easier to simplify your fraction with the brackets in place.

3 To DIVIDE fractions:

1. Change the second fraction to its reciprocal.

2. Change $\div$ to $\times$.

3. Multiply the fractions.

4. Simplify if possible.

$$\frac{x^2}{3} \div \frac{x}{6} = \frac{x^2}{3} \times \frac{6}{x} = \frac{{}^{2}6x^2}{{}_{1}3x} = 2x$$

To find the reciprocal of a fraction you turn it upside down.

Worked example

grade A*

Simplify fully $\dfrac{x^2 - 25}{x^2 + 7x + 10}$

$$\frac{x^2 - 25}{x^2 + 7x + 10} = \frac{(x+5)(x-5)}{(x+5)(x+2)}$$
$$= \frac{x-5}{x+2}$$

EXAM ALERT!

Only one in ten students got full marks on this question. The trick is to factorise the top and the bottom before simplifying. Remember $a^2 - b^2 = (a+b)(a-b)$

This was a real exam question that caught students out – **be prepared!**

Now try this

grade A

edexcel

grade A*

1. Simplify fully
$$\frac{x+3}{4} + \frac{x-5}{3} \quad \textbf{(3 marks)}$$

2. Simplify fully

(a) $\dfrac{x^2 - 3x}{x^2 - 8x + 15}$ (b) $\dfrac{2x^2 - x - 3}{4x^2 - 9}$ **(6 marks)**

Proof

You can use algebra to PROVE facts about numbers.

Using algebra helps you to prove that something is true for EVERY number.

In a proof question the working IS the answer.

If the question says 'SHOW THAT...' or 'PROVE THAT...', you need to write down every stage of your working.

Golden rule

If you need to prove something about numbers then you always use algebra.

Algebraic proof toolkit

Use n to represent any whole number.

Number fact	Written using algebra
Even number	$2n$
Odd number	$2n + 1$ or $2n - 1$
Multiple of 3	$3n$
Consecutive numbers	$n, n + 1, n + 2, \ldots$
Consecutive even numbers	$2n, 2n + 2, 2n + 4, \ldots$
Consecutive odd numbers	$2n + 1, 2n + 3, 2n + 5, \ldots$
Consecutive square numbers	$n^2, (n + 1)^2, (n + 2)^2, \ldots$

Worked example

grade A*

Prove that the sum of three consecutive integers is always divisible by 3

$n, n + 1$ and $n + 2$ represent any three consecutive integers.
$$n + (n + 1) + (n + 2) = 3n + 3$$
$$= 3(n + 1)$$
$n + 1$ is an integer, so $3(n + 1)$ is divisible by 3.

To **prove** the statement you need to show that it is true for **any** three consecutive integers. You can do this using algebra.
1. Write the first integer as n and the next two integers as $n + 1$ and $n + 2$.
2. Write an expression for the sum of your three integers. Brackets can help to make your working clearer. Simplify your expression.
3. Factorise the expression.
4. Explain why the final expression is divisible by 3.

True or false?

You can only prove things that are true.

It's easy to explain why something isn't true.

To prove that the statement below isn't true you just need to write down one COUNTER-EXAMPLE.

Statement: The sum of any two prime numbers is an even number.

Counter-example: 2 and 3 are prime numbers.

$2 + 3 = 5$, which is not an even number.

A counter-example shows that a statement is false.

Worked example

grade A*

Show that $(n - 1)^2 + (n + 1)^2 = 2(n^2 + 1)$

$(n - 1)^2 + (n + 1)^2$
$= (n - 1)(n - 1) + (n + 1)(n + 1)$
$= n^2 - 2n + 1 + n^2 + 2n + 1$
$= 2n^2 + 2$
$= 2(n^2 + 1)$

Start with the expression on the left-hand side. Use multiplying out, simplifying and factorising to work towards the expression on the right-hand side.

Now try this

 edexcel

grade A*

Prove that the difference between the squares of any two consecutive odd integers is always divisible by 4   **(4 marks)**

Start by writing down two odd integers in terms of n. Then set up the expression needed.
Remember that $(2n + 1)^2 \neq 4n^2 + 1$.

Problem-solving practice

About half of the questions on your exam will need problem-solving skills.

These skills are sometimes called AO2 and AO3.

Practise using the questions on the next two pages.

For these questions you might need to:

- choose which mathematical technique or skill to use
- apply a technique in a new context
- plan your strategy to solve a longer problem
- show your working clearly and give reasons for your answers.

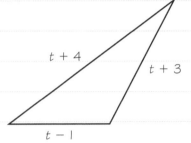

The perimeter of this triangle is 19 cm.

All lengths on the diagram are in cm.

Work out the value of t.   (4 marks)

Linear equations 2 p. 24

Use the information in the question to write an equation. Solve your equation to work out the value of t.

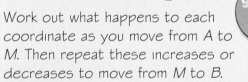 **grade D**

TOP TIP

Remember to simplify your equation by collecting like terms before solving.

 AB is a line segment.
A is the point (5, 2, −4).

The midpoint M of the line segment AB has coordinates (−1, −4, 2).

Jim says that B has coordinates (2, −1, −1).

Jim is wrong. Explain why.   (2 marks)

3-D coordinates p. 27

Work out what happens to each coordinate as you move from A to M. Then repeat these increases or decreases to move from M to B.

grade B

TOP TIP

Draw a sketch to make sure you know what you need to work out.

---

③ A straight line has equation
$2y - 6x = 5$

The point $(k, 6)$ lies on the line.

Find the value of k.   (2 marks)

Straight-line graphs p. 25

Substitute $x = k$ and $y = 6$ into the equation of the line. This will give you an equation with one unknown value. Solve the equation to find the value of k.

grade A

TOP TIP

You can substitute the x- and y-values of a point into the equation of a graph. If the values make the equation true, the point with those coordinates lies on the graph.

Problem-solving practice

4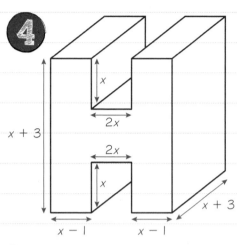

Diagram **NOT** accurately drawn

The diagram shows a prism.

All measurements are in cm.

All corners are right angles.

The volume of the prism is V cm³.

Find a formula for V.

Simplify your answer. **(4 marks)**

Formulae p. 29
Prisms p. 56

A good way to do this is to divide the H-shape up into smaller parts. Label the parts A, B, C, etc. and write down an expression for the area of each part. Add your expressions together and simplify if you can. Remember that the volume of a prism is the area of the cross-section multiplied by the length.

TOP TIP

Don't panic if a question gives lengths as letters. Follow the same steps as you would if the lengths were given as numbers and be really careful when simplifying any expressions.

5

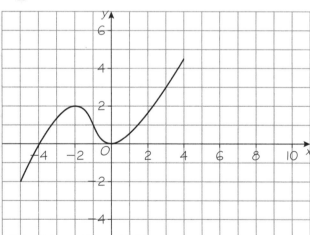

The graph of $y = f(x)$ is shown on the grid.

(a) On the grid sketch the graph of
$y = f(x) + 2$ **(2 marks)**

(b) Fully describe the transformation that would map the graph of $y = f(x)$ onto the graph of $y = -f(x)$ **(2 marks)**

Transformations 1 p. 45

A 'sketch' graph still needs to be drawn accurately with a sharp pencil. First draw crosses for the key points that the graph should pass through. Make sure your curve passes through all of those points.

grade
A*

TOP TIP

You can check your answers. The point (3, 3) lies on the graph of $y = f(x)$ so you know that $f(3) = 3$. This means that (3, 5) should lie on the graph of $y = f(x) + 2$.

Angle properties

A*
A
B
C
D

You need to remember all of these angle properties and their correct names.

CORRESPONDING
ANGLES are equal.

ALTERNATE ANGLES
are equal.

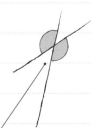

ALLIED ANGLES
add up to 180°.

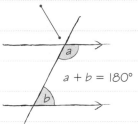

$a + b = 180°$

Parallel lines are
marked with arrows.

OPPOSITE ANGLES are equal.

These are useful angle facts for triangles
and parallelograms:

Interior
angle

b

a

Exterior
angle

$a + b$

The exterior angle of a
triangle is equal to the
sum of the interior angles
at the other two vertices.

The opposite
angles of a
parallelogram
are equal.

You need to know the proofs of the angle
properties of triangles and quadrilaterals.

Golden rule

When answering angle problems, you
need to give a reason for each step
of your working.

Geometric proof checklist

To prove a geometric fact you need to:
* write down each step of
  your working clearly ✓
* give a reason for each
  step of your working ✓
* use the correct words for your
  reasons. ✓

Worked example grade **D**

PQRS is a parallelogram.
(a) Give a reason why angle
   a is the same size as
   angle d.

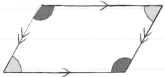

a and d are alternate angles.
Alternate angles are equal.

(b) Prove that the opposite angles in a
    parallelogram are equal.
$a = d$ (alternate angles are equal)
$b = c$ (alternate angles are equal)
$a + b = d + c$
So the opposite angles are equal.

Now try this

edexcel

ABCD is a rhombus.
Prove that *AC* bisects
angle *DCB*. **(3 marks)**

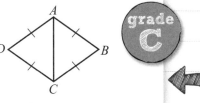

grade **C**

You need to show that angle *DAC* is the
same as angle *BAC*. You could start by
writing angle *BAC* as *x*. If you say that
any other angle is also equal to *x* make
sure that you give a reason.

A*
A
B
C
D

Solving angle problems

Angle problems come up every year. You need to be confident in them if you want a C or D. Write down reasons for every step of your working.

Worked example

grade
D

Work out the size of the angle marked *x*.
Give reasons for each step of your working.

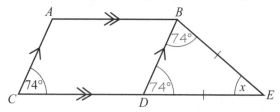

∠*BDE* = 74° (corresponding angles are equal)

∠*DBE* = 74° (base angles in an isosceles triangle are equal)

x + 74° + 74° = 180° (angles in a triangle add up to 180°)

x = 180° − 148°

x = 32°

Use the properties on the diagram:
AB is parallel to *CD*
AC is parallel to *BD*
BE is equal to *DE*

Worked example

grade
D

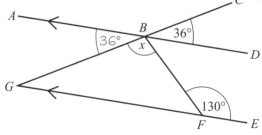

ABD and *GFE* are parallel lines.
GBC is a straight line.

Work out the size of the angle marked *x*.

∠*ABG* = 36° (opposite angles are equal)

∠*ABF* = 130° (alternate angles are equal)

x = 130° − 36°

x = 94°

You can use the angle symbol (∠) and the triangle symbol (△) to show your working. Write any angles you can work out on the diagram. Every time you write down an angle you should write down a reason.

Reasons

Use these reasons in angle problems.
* Angles on a straight line add up to 180°.
* Angles around a point add up to 360°.
* Opposite angles are equal.
* Corresponding angles are equal.
* Alternate angles are equal.
* Angles in a triangle add up to 180°.
* Angles in a quadrilateral add up to 360°.
* Base angles of an isosceles triangle are equal.

Now try this

edexcel

grade
D

ABC and *DEFG* are straight lines.
AC is parallel to *DG*.
BE = *BF*.
Angle *ABE* = 62°.

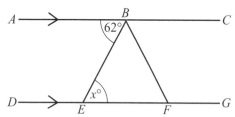

Diagram **NOT** accurately drawn

(a) (i) Find the value of *x*. **(1 mark)**
     (ii) Give a reason for your answer. **(1 mark)**
(b) Explain why angle *EBF* is 56°. **(2 marks)**

If a triangle has two equal sides, it is isosceles.

Use the fact that the angles in a triangle add up to 180° to write an equation.

Solve your equation to find the size of angle *x*.

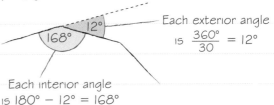

A*
A
B
C
D

Angles in polygons

Polygon questions are all about interior and exterior angles.

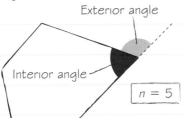

Exterior angle

Interior angle

$n = 5$

Use these formulae for a polygon with n sides.

Sum of interior angles $= 180° \times (n - 2)$

Sum of exterior angles $= 360°$

This diagram shows part of a regular polygon with 30 sides.

168°   12°

Each exterior angle
is $\frac{360°}{30} = 12°$

Each interior angle
is $180° - 12° = 168°$

Don't try to draw a 30-sided polygon!

If there's no diagram given in a polygon question, you probably don't need to draw one.

Regular polygons

In a regular polygon all the sides are equal and all the angles are equal.

If a regular polygon has n sides then each exterior angle is $\dfrac{360°}{n}$

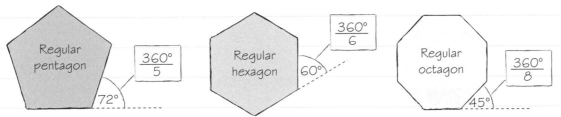

Regular pentagon $\dfrac{360°}{5}$ 72°

Regular hexagon $\dfrac{360°}{6}$ 60°

Regular octagon $\dfrac{360°}{8}$ 45°

You can use the fact that the angles on a straight line add up to 180° to work out the size of one of the interior angles.

Worked example

grade D

Work out the size of an exterior angle of a regular pentagon.

Exterior angles of polygon add up to 360°
So exterior angle is $360° \div 5 = 72°$

EXAM ALERT!

Two-thirds of students dropped a mark here.

A pentagon has 5 sides. You need to know the names of the polygons with 3 to 8 sides (triangle, quadrilateral, pentagon, hexagon, heptagon, octagon) and 10 sides (decagon) for your exam.

This was a real exam question that caught students out – **be prepared!**

 Results**Plus**

Now try this

edexcel

 grade D

(a) The size of each exterior angle of a regular polygon is 40°.
Work out the number of sides of the regular polygon.   **(2 marks)**

 grade C

(b) *ABCDEF* is a regular hexagon.
BAG and *EFG* are straight lines.
Work out the size of angle *AGF*.
Give your reason for your answer.   **(3 marks)**

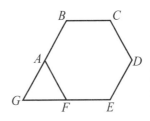

Diagram **NOT** accurately drawn

Plan and elevation

Plans and elevations are 2-D drawings of 3-D shapes as seen from different directions.

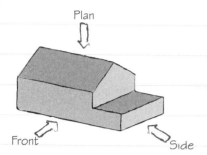

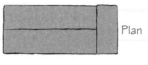

Plan

The PLAN is the view from above.

Front elevation

The FRONT ELEVATION is the view from the front.

Side elevation

This line shows a change in depth.

The SIDE ELEVATION is the view from the side.

You might be asked to draw a 3-D shape on isometric paper.

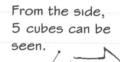

From above, 4 cubes can be seen.

Plan

From the side, 5 cubes can be seen.

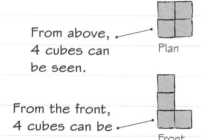

From the front, 4 cubes can be seen.

Front elevation   Side elevation

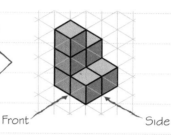

There are a total of 7 cubes in this shape.

Front   Side

Worked example

The diagram shows a solid shape.

On the grid below draw a plan, front and side elevations of the shape.

Front   Side

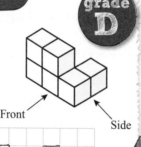

Plan   Front elevation   Side elevation

grade D

Imagine tracing an image of the shape on each side of a box.
Unfold the box to get your plan and elevations.

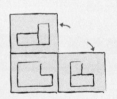

Put lines within the plan and side elevation to show where there is a change in height or depth.

The vertical lines on the plan show where the height of the shape changes.

Now try this

Here are the plan and front elevation of a prism. The front elevation shows the cross-section of the prism.

(a) On a grid, draw a side elevation of the prism.   **(3 marks)**

(b) Draw a 3-D sketch of the prism.   **(2 marks)**

edexcel

grade C

Plan

Front elevation

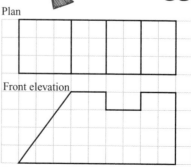

Perimeter and area

Triangle

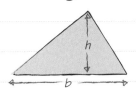

Area = $\frac{1}{2}bh$

Learn this formula ✓

Parallelogram

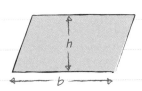

Area = bh

Learn this formula ✓

Trapezium

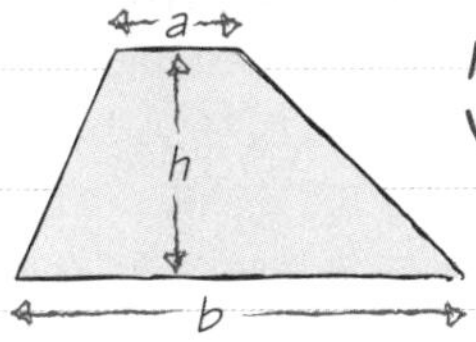

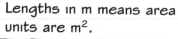

h is the VERTICAL HEIGHT

Area = $\frac{1}{2}(a + b)h$

Given on the formula sheet ✓

You can calculate areas and perimeters of more complex shapes by splitting them into parts.

You might need to draw some extra lines on your diagram and add or subtract areas.

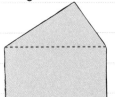
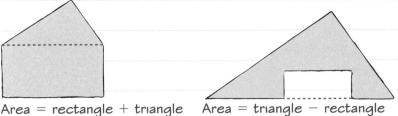

Area = rectangle + triangle   Area = triangle − rectangle

Area basics

Lengths are all in the same units. ✓

Give units with the answer. ✓

Lengths in cm means area units are cm². ✓

No calculator. ✓

Lengths in m means area units are m². ✓

Worked example

grade **D**

The diagram shows a garden bed. Adrian wants to cover the bed with grass seed. A packet of grass seed will cover 10 m².

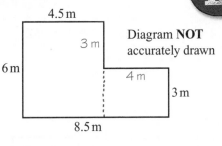

Diagram **NOT** accurately drawn

(a) How many packets of grass seed does Adrian need to buy?

Area = 6 × 4.5 + 4 × 3 = 27 + 12 = 39 m²
Adrian needs to buy 4 packets of grass seed.

Adrian also wants to build a fence around the edge of the garden bed.

(b) Calculate the total length of Adrian's fence.

6 + 4.5 + 3 + 4 + 3 + 8.5 = 29 m

Draw a dotted line to divide the diagram into two rectangles.

You have to use the information in the question to work out the missing lengths. The diagram is **not accurately drawn,** so you can't use a ruler to measure.

8.5 m − 4.5 m = 4 m

6 m − 3 m = 3 m

Write these lengths on your diagram.

Make sure you answer the question that has been asked.

You need to say how many packets of grass seed Adrian needs to buy.

Now try this

The diagram shows a field.

A farmer wants to put some sheep in the field.

He decides that each sheep needs at least 5 m² for grazing.

Work out the maximum number of sheep he will put in this field.   **(4 marks)**

Diagram **NOT** accurately drawn

edexcel

grade **C**

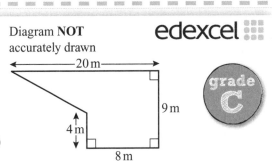

A* A A B C D

Prisms

A prism is a 3-D shape with a constant CROSS-SECTION. If the cross-section is a rectangle then we call the prism a cuboid.

Cuboid

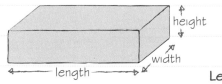

height
width
← length →

Volume = length × width × height   **Learn this formula ✓**

Prism

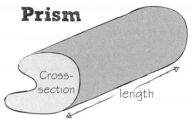

Cross-section   length

Volume = area of cross-section × length   **Given on the formula sheet ✓**

You might need to work out the area of the cross-section before working out the volume of the prism.

You should draw a sketch of the cross-section separately to work out its area.

6 cm   4 cm   7 cm   10 cm   →   ↕2 cm   ↕4 cm   ←7 cm→

Area of cross-section = area of rectangle + area of triangle
$$= 7 \times 4 + \tfrac{1}{2} \times 7 \times 2 = 35\,\text{cm}^2$$
Volume of prism = 35 × 10 = 350 cm³

Surface area

To work out the surface area of a 3-D shape, you need to add together the areas of all the faces. It's a good idea to sketch each face with its dimensions. Remember to include the faces that you can't see.

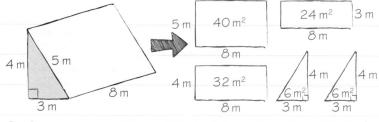

4 m   5 m   8 m   3 m   →

5 m   40 m²   8 m
4 m   32 m²   8 m
24 m²   3 m   8 m
4 m   6 m²   3 m   4 m   6 m²   3 m

Surface area = 40 + 32 + 24 + 6 + 6 = 108 m²

A light bulb box measures 8 cm by 8 cm by 10 cm.

Light bulb boxes are packed into cartons.

A carton measures 40 cm by 40 cm by 60 cm.

Work out the number of light bulb boxes which can completely fill one carton.

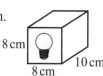

8 cm   10 cm   8 cm

40 cm   Carton   60 cm   40 cm

Diagram **NOT** accurately drawn

Volume of carton = 40 × 40 × 60 cm³   Volume of light bulb box = 8 × 8 × 10 cm³

Number of light bulb boxes which can fill one carton = $\dfrac{\overset{5}{40} \times \overset{5}{40} \times \overset{6}{60}}{8_1 \times 8_1 \times 10_1} = 5 \times 5 \times 6 = 150$

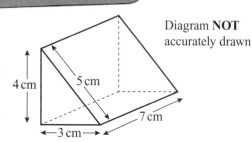

4 cm   5 cm   7 cm   ←3 cm→

Diagram **NOT** accurately drawn

edexcel

(a) Show that the total surface area of the prism is 96 cm². **(3 marks)**   grade **D**

(b) Calculate the volume of the triangular prism. **(4 marks)**   grade **C**

Circles and cylinders

A* A B C D

You need to learn these formulae for circles and cylinders. They're not given on the formula sheet.

Circle

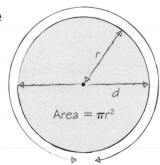

Circumference = $2\pi r$
= πd

Area = πr^2

Cylinder

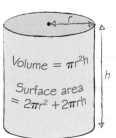

Volume = $\pi r^2 h$

Surface area
= $2\pi r^2 + 2\pi rh$

Worked example

grade **C**

Here is a tile in the shape of a semicircle.

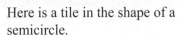

8 cm

The diameter of the semicircle is 8 cm.
Work out the perimeter of the tile.
Give your answer correct to 2 decimal places.

Perimeter = $(\frac{1}{2} \times \pi \times 8) + 8$
= 20.5663...
= 20.57 cm (to 2 d.p.)

EXAM ALERT!

Only 30% of students got full marks on this question.

Do **not** use the formula for the area of a circle to work out a perimeter.

Perimeter = length of arc + diameter of semicircle

The length of the arc is equal to **half** the circumference.

Write down at least four figures after the decimal point on the calculator display before giving your final answer.

In terms of π

If a question asks for an EXACT VALUE or an answer IN TERMS OF π then don't use the π button on your calculator. Write your answer as a whole number or fraction multiplied by π.

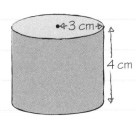

3 cm

4 cm

Volume of cylinder = $\pi r^2 h$
= $\pi \times 3^2 \times 4$

EXACT ANSWER
Volume = 36π cm^3

ROUNDED ANSWER
Volume = 113 cm^3 (to 3 s.f.)

Now try this

 edexcel

The top of a table is a circle.
The radius of the top of the table is 50 cm.
(a) Work out the area of the top of the table.   **(2 marks)**

grade **D**

The base of the table is a circle.
The diameter of the base of the table is 40 cm.
(b) Work out the circumference of the base of the table.   **(2 marks)**

A*
A
B
C
D

Sectors of circles

You can find the area of a sector by working out what fraction it is of the whole circle.

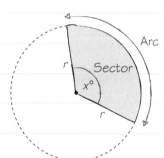

For a sector with angle $x°$ of a circle with radius r:

Sector $= \dfrac{x}{360}$ of the whole circle so

Area of sector $= \dfrac{x}{360} \times \pi r^2$

Arc length $= \dfrac{x}{360} \times 2\pi r$

Learn these formulae. ✓

You can give answers in terms of π. ✓

Worked example

grade A

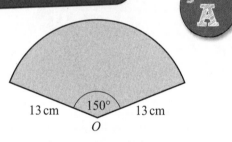

13 cm   150°   13 cm

O

The diagram shows a wooden game piece in the shape of a sector of a circle, centre O.

Calculate the <u>perimeter</u> of the game piece.
Give your answer correct to 3 significant figures.

Arc length $= \dfrac{x}{360} \times 2\pi r$

$= \dfrac{150}{360} \times 2\pi \times 13$

Perimeter $=$ arc length $+$ radius $+$ radius

$= \dfrac{150}{360} \times 2\pi \times 13 + 13 + 13$

$= 60.0339... = 60.0$ cm (to 3 s.f.)

Finding a missing angle

You can use the formulae for arc length or area to find a missing angle in a sector. Practise this method if you're going for an A*.

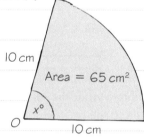

10 cm

Area $= 65$ cm²

$x°$

O 10 cm

Area of sector $= \dfrac{x}{360} \times \pi r^2$

$65 = \dfrac{x}{360} \times \pi(10)^2$

$x = \dfrac{65 \times 360}{\pi(10)^2}$

$ = 74.4845...$

$ = 74.5°$ (to 3 s.f.)

Don't round your values until the end of the calculation.

Now try this

edexcel

grade A

The diagram shows a sector of a circle, centre O.
The radius of the circle is 6 cm.
The angle of the sector is 120°.
Calculate the area of the sector.
Give your answer correct to 3 significant figures. **(2 marks)**

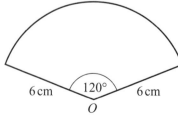

6 cm   120°   6 cm

O

Diagram **NOT** accurately drawn

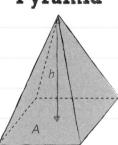

A* A B C D

Volumes of 3-D shapes

Cuboid

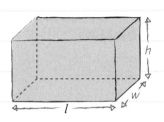

Volume of cuboid
= length × width × height
= lwh

Cylinder

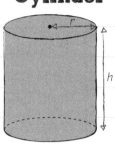

Volume of cylinder
= area of base × height
= $\pi r^2 h$

Pyramid

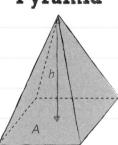

Volume of pyramid
= $\frac{1}{3}$ × area of base × vertical height
= $\frac{1}{3}Ah$

Learn these volume formulae. ✓

Cone

Volume of cone
= $\frac{1}{3}$ × area of base × vertical height
= $\frac{1}{3}\pi r^2 h$

Sphere

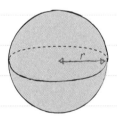

Volume of sphere = $\frac{4}{3}\pi r^3$

These volume formulae are on the formula sheet. ✓

Worked example

grade **A**

A cone has a base radius of 6 cm and a height of 12.5 cm.

Work out the volume of the cone.

Give your <u>answer in terms of π.</u>

12.5 cm

6 cm

Volume = $\frac{1}{3}$ × (π × 6^2) × 12.5
= $\frac{1}{3}$ × 36π × 12.5
= 150π cm^3

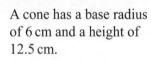

Area of circular base = πr^2

To work out the volume of a composite shape, work out the volumes of the shapes it is made from and add the volumes together.

Now try this

edexcel ▦

The diagram shows a storage tank.

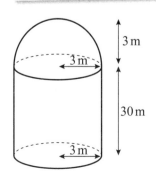

3 m

3 m

30 m

3 m

The storage tank consists of a hemisphere on top of a cylinder.

The height of the cylinder is 30 metres.

The radius of the cylinder is 3 metres.

The radius of the hemisphere is 3 metres.

Calculate the total volume of the storage tank.

Give your answer correct to 3 significant figures.

(3 marks)

grade **A***

-A*-
-A-
-B-
-C-
-D-

Pythagoras' theorem

Pythagoras' theorem is a really useful rule. You can use it to find the length of a missing side in a right-angled triangle.

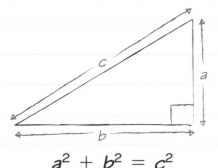

$$a^2 + b^2 = c^2$$

Pythagoras checklist

short^2 + short^2 = long^2 ✓

Right-angled triangle. ✓

Lengths of two sides known. ✓

Length of third side missing. ✓

Learn this. It's not on the formula sheet. ✓

Worked example

grade C

Work out the length of the missing side.

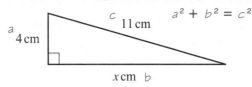

$$4^2 + x^2 = 11^2$$
$$x^2 = 11^2 - 4^2$$
$$= 121 - 16 = 105$$
$$x = \sqrt{105} = 10.2 \text{ (to 1 d.p.)}$$

So $x = 10.2$ cm

Be really careful when the missing length is one of the **shorter** sides.
1. Label the longest side of the triangle c.
2. Label the other two sides.
3. Substitute the values into the formula.
4. Rearrange and solve.
5. Write units with your answer.

Pythagoras questions come in lots of different forms. Just look for the right-angled triangle.

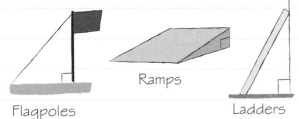

Flagpoles   Ramps   Ladders

Calculator skills

Use these buttons to find squares and square roots with your calculator.

You might need to use the S⇔D key to get your answer as a decimal number.

Now try this

grade C

edexcel

Calculate the area of this right-angled triangle.

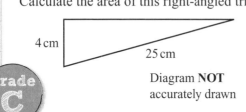

Diagram **NOT** accurately drawn

(3 marks)

You need to find the length of the missing side before you can find the area of the triangle. You know the lengths of the other two sides and the triangle is right-angled, so you can use Pythagoras' theorem. For a reminder about areas of triangles have a look at page 55.

Surface area

Cone

The formula for the CURVED SURFACE AREA of a cone is given on the formula sheet.

> **Curved surface area of cone** $= \pi r l$
>
>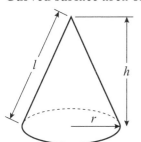

Be careful! This formula uses the slant height, l, of the cone.

To calculate the TOTAL surface area of the cone you need to add the area of the base. Surface area of cone $= \pi r^2 + \pi r l$

Sphere

The formula for the surface area of a sphere is given on the formula sheet.

> **Surface area of sphere** $= 4\pi r^2$
>
>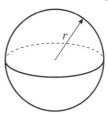

For a reminder about areas of circles and surface areas of cylinders have a look at page 57.

A hemisphere is half a sphere, so the area of the curved surface is $\frac{1}{2} \times 4\pi r^2$.

Worked example

 grade A

The diagram shows a cone with vertical height 12 cm and base diameter 10 cm.

Calculate the curved surface area of the cone.

Give your answer in terms of π.

[diagram: cone with 12 cm height, 10 cm base]

$r = 5$

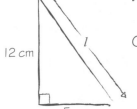

$l^2 = 12^2 + 5^2 = 169$

$l = 13 \text{ cm}$

Curved surface area
$= \pi r l$
$= \pi \times 13 \times 5$
$= 65\pi \text{ cm}^2$

To work out the curved surface area you need to know the radius and the slant height. You are given the **diameter** and the **vertical height**.

The radius is half the diameter $= 5$ cm.

To calculate the slant height you need to use Pythagoras' theorem. Sketch the right-angled triangle containing the missing length.

There is more about Pythagoras' theorem on page 60.

Remember to leave your final answer in terms of π.

Compound shapes

You can calculate the surface area of more complicated shapes by adding together the surface area of each part.

[diagram: cylinder with hemispherical ends, 4 cm and 6 cm]

Surface area $= \pi(4)^2 + 2\pi(4)(6) + \frac{1}{2}[4\pi(4)^2]$
$= 96\pi \text{ cm}^2$

Now try this

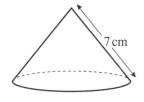

The curved surface area of this cone is $35\pi \text{ cm}^2$.

(a) Write down the radius of the cone. **(2 marks)**

(b) Work out the area of the base of the cone.
Give your answer correct to 3 significant figures.
(2 marks)

edexcel ⠿

grade A

**A*
A
B
C
D**

Converting units

You can convert between METRIC UNITS by multiplying or dividing by 10, 100 or 1000.

Length

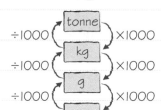

Weight

÷1000 (tonne) ×1000
÷1000 (kg) ×1000
÷1000 (g) ×1000
(mg)

Volume or capacity

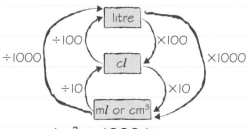

$1 m^3 = 1000$ litres

Imperial units

You need to remember these conversions for your exam.

Metric unit	Imperial unit
1 (kg)	2.2 pounds (lb)
1 litre (*l*)	$1\frac{3}{4}$ pints
4.5 litres	1 gallon
8 km	5 miles
30 cm	1 foot (ft)

When converting between imperial units you will be GIVEN the conversions.

Converting compound units

To convert between measures of speed you need to convert one unit first then the other. Write the new units at each step of your working. To convert 72 km/h into m/s:

72 km/h → 72 × 1000 = 72 000 m/h
72 000 m/h → 72 000 ÷ 3600 = 20 m/s
1 hour = 60 × 60 = 3600 seconds

Worked example   grade C

Waseem drives at an average speed of 60 mph. How long will it take him to drive 120 km?

8 km = 5 miles
60 ÷ 5 = 12
8 × 12 = 96
60 mph = 96 km/h

Time = $\dfrac{distance}{average\ speed}$

$= \dfrac{120}{96} = 1\frac{1}{4}$ hours

First of all, you need to convert the speed from mph to km/h.

Convert 60 miles into km. You can use equivalent ratios:

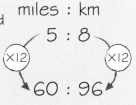

miles : km
5 : 8
(×12)   (×12)
60 : 96

Kilometres are smaller than miles so the speed in km/h should be a larger number than the speed in mph.

Then use the formula triangle for speed (see page 64) to work out the answer.

Now try this   grade C

edexcel

Jane walks at an average speed of 5 km/h.
Mattie walks at an average speed of 3 miles per hour.
How long will they each take to walk 5 miles? **(3 marks)**

Speed is a compound measure. Another compound measure is density. Density is covered on page 65.

Units of area and volume

A*
A
B
C
D

In your exam, you may be asked to convert between different units of area and volume.

These two squares have the same area.

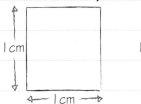
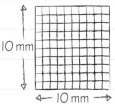

So 1 cm² = 100 mm².

These two cubes have the same volume.

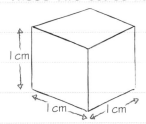
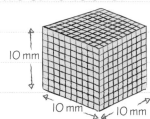

So 1 cm³ = 1000 mm³.

Area conversions

$1 \text{ cm}^2 = 10^2 \text{ mm}^2 = 100 \text{ mm}^2$

$1 \text{ m}^2 = 100^2 \text{ cm}^2 = 10\,000 \text{ cm}^2$

$1 \text{ km}^2 = 1000^2 \text{ m}^2 = 1\,000\,000 \text{ m}^2$

Volume conversions

$1 \text{ cm}^3 = 10^3 \text{ mm}^3 = 1000 \text{ mm}^3$

$1 \text{ m}^3 = 100^3 \text{ cm}^3 = 1\,000\,000 \text{ cm}^3$

$1 \text{ litre} = 1000 \text{ cm}^3$

$1 \text{ m}l = 1 \text{ cm}^3$

There is more on converting metric units on page 62.

Worked example

 grade C

Change 125 cm³ into mm³.

$125 \times 10^3 = 125\,000$

$125 \text{ cm}^3 = 125\,000 \text{ mm}^3$

EXAM ALERT!

9 out of 10 students got this question wrong.

To convert from cm into mm you multiply by 10. So to convert from cm³ into mm³ you multiply by 10³, or 1000.

This was a real exam question that caught students out – **be prepared!**   Results Plus

Unit conversion checklist

When converting to a bigger unit, divide by a power of 10.   ✓

When converting to a smaller unit, multiply by a power of 10.   ✓

The multiplier for an area conversion is the length multiplier squared.   ✓

The multiplier for a volume conversion is the length multiplier cubed.   ✓

Work out the volume of the tank. Make sure your answer is in cm³.

Use 1 litre = 1000 cm³ to convert your answer into litres.

 grade C

Now try this

1. (a) Change 2.5 m² to cm².   **(1 mark)**

   (b) Change 8560 mm² to cm².   **(1 mark)**

   (c) Change 7 cm³ to mm³.   **(1 mark)**

   (d) Change 856 000 cm³ to m³.   **(1 mark)**

2. The diagram shows a tank in the shape of a cuboid.
   Work out how many litres of water the tank can hold.

 grade B

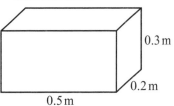

0.3 m

Diagram **NOT** accurately drawn

0.2 m

0.5 m

(4 marks)

edexcel

Speed

A*
A
B
C
D

This is the formula triangle for speed.

Average speed ⋅   ⋅ Distance

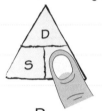

D
S T

⋅ Time

$$\text{Average speed} = \frac{\text{total distance travelled}}{\text{total time taken}}$$

$$\text{Time} = \frac{\text{distance}}{\text{average speed}}$$

Distance = average speed × time

Using a formula triangle

Cover up the quantity you want to find with your finger.

The position of the other two quantities tells you the formula.

$$T = \frac{D}{S} \qquad S = \frac{D}{T} \qquad D = S \times T$$

Units

The most common units of speed are

• metres per second: m/s
• kilometres per hour: km/h
• miles per hour: mph

The units in your answer will depend on the units you use in the formula.

When distance is measured in km and time is measured in hours, speed will be measured in km/h.

When you are calculating a distance or time, you MUST make sure that the units of the other quantities match.

Minutes and hours

For questions on speed, you need to be able to convert between minutes and hours.

Remember there are 60 minutes in 1 hour.

To convert from minutes to hours you divide by 60.

24 minutes = 0.4 hours $\frac{24}{60} = \frac{2}{5} = 0.4$

To convert from hours to minutes you multiply by 60.

3.2 hours = 192 minutes   3.2 × 60 = 192
= 3 hours 12 minutes

Worked example

grade **C**

A plane travels at a constant speed of 600 km/h for 45 minutes.
How far has it travelled?

45 minutes = $\frac{45}{60}$ hours = $\frac{3}{4}$ hour

D = S × T
  = 600 × $\frac{3}{4}$ = $\frac{600 \times 3}{4}$ = $\frac{1800}{4}$ = 450

The plane has travelled 450 km.

D
S T

Speed checklist

Draw formula triangle.   ✓
Make sure units match.   ✓
Give units with answer.   ✓

The journey time in part (b) needs to be changed into hours.
Any minutes calculated must be changed to a fraction of an hour in its simplest form before calculating the speed.

Now try this

(a) Daniel leaves his house at 07:00.
grade **C**   He drives 100 miles to a meeting. He drives at
an average speed of 40 miles per hour.
At what time does Daniel arrive at the
meeting?   **(2 marks)**

(b) Daniel leaves the meeting at 17:20.
He drives 100 miles back home.
He arrives home at 20:00.
What was Daniel's average speed?
(2 marks)   grade **C**

edexcel ⠿

Density

The density of a material is its mass per unit volume.

This is the formula triangle for density.

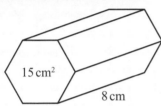

$$\text{Density} = \frac{\text{mass}}{\text{volume}}$$

$$\text{Volume} = \frac{\text{mass}}{\text{density}}$$

$$\text{Mass} = \text{density} \times \text{volume}$$

For a reminder of how to use a formula triangle have a look at page 64.

- A*
- A
- B
- C
- D

Worked example

grade **C**

15 cm²

8 cm

The diagram shows a solid hexagonal prism.

The area of the cross-section of the prism is 15 cm².

The length of the prism is 8 cm.

The prism is made from wood with a density of 0.8 grams per cm³.

Work out the mass of the prism.

Volume of prism
= area of cross-section × length
= 15 × 8
= 120 cm³

$M = D \times V$
= 0.8 × 120
= 96

The mass of the prism is 96 grams.

Units

The most common units of density are

- grams per cubic centimetre: g/cm³
- kilograms per cubic metre: kg/m³

Mass = density × volume

You are given the density so you need to work out the volume of the prism.

The formula for the volume of a prism is given on the formula sheet.

The density is in grams per cm³ and the volume is in cm³ so the mass will be in grams.

Worked example

grade **C**

The mass of 5 m³ of copper is 44 800 kg. Work out the density of copper.

$$D = \frac{M}{V} = \frac{44\,800}{5} = 8960 \text{ kg/m}^3$$

Mass is given in kg and volume is given in m³ so the units of density will be kg/m³.

Now try this

edexcel

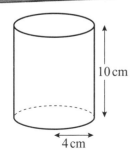

10 cm

4 cm

Diagram **NOT** accurately drawn

A solid cylinder is made from wood. It has a radius of 4 cm and a height of 10 cm.
The density of the wood is 0.6 grams per cm³.
Work out the mass of the cylinder.
Give your answer correct to 3 significant figures.   **(4 marks)**

grade **B**

-A*-
-A-
-B-
-C-
-D-

Congruent triangles

Two triangles are CONGRUENT if they have exactly the same shape and size.

To prove this you have to show that ONE of these four conditions is true.

1 **SSS** (three sides are equal)

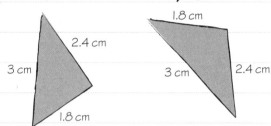

2.4 cm
3 cm
1.8 cm

1.8 cm
3 cm   2.4 cm

2 **AAS** (two angles and a corresponding side are equal)

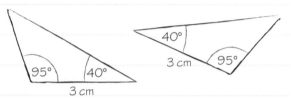

95° 40°
3 cm

40°
3 cm 95°

3 **SAS** (two sides and the included angle are equal)

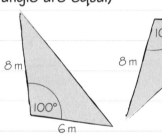

8 m
100°
6 m

6 m
100°
8 m

The angle must be BETWEEN the two sides for SAS.

4 **RHS** (right angle, hypotenuse and a side are equal)

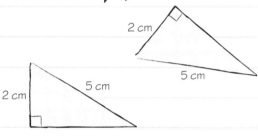

2 cm
5 cm

2 cm
5 cm

Worked example

grade A*

ABC is an equilateral triangle.
D lies on *BC*.
AD is perpendicular to *BC*.
Prove that triangle *ADC* is congruent to triangle *ADB*.

A
B   D   C

∠ADB = ∠ADC = 90°
(AD is perpendicular to BC)
AD = AD
(It is a common side on both triangles)
∠ABD = ∠ACD
(Base angles of an isosceles triangle are equal)
So △ADC is congruent to △ADB. (AAS)

EXAM ALERT!

Congruent triangle questions are often tricky. Only 5% of students got this question right in the exam.

'Prove that…' means you need to write down a reason for every step of your working.

You need to show that one of the conditions for congruence is true.

You can **only** use the properties given in the question to prove congruence.

Write down three things that are equal, give a reason for each and write the condition for congruence at the end. The condition here is AAS.

This was a real exam question that caught students out – **be prepared!**   ResultsPlus

---

Now try this

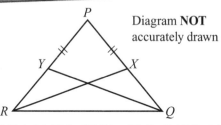

P
Y   X
R   Q

Diagram **NOT** accurately drawn

Triangle *PQR* is isosceles with *PQ* = *PR*.
X is a point on *PQ*.
Y is a point on *PR*.
PX = *PY*.
Prove that triangle *PQY* is congruent to triangle *PRX*.   **(3 marks)**

grade A*

edexcel

Similar shapes 1

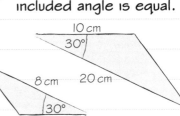
-A*-
-A-
-B-
-C-
-D-

Shapes are SIMILAR if one shape is an enlargement of the other.

SIMILAR TRIANGLES satisfy one of these three conditions:

 All three pairs of angles are equal.

 All three pairs of sides are in the same ratio.

 Two sides are in the same ratio and the included angle is equal.

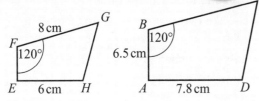

You might need to use a geometric proof to prove that two triangles are similar.

Worked example

grade **A**

The diagram shows two mathematically similar shapes.

(a) Calculate the length *BC*.

$$\frac{BC}{FG} = \frac{AD}{EH}$$

$$\frac{BC}{8} = \frac{7.8}{6}$$

$$BC = \frac{7.8 \times 8}{6} = 10.4 \text{ cm}$$

(b) Calculate the length *EF*.

$$\frac{EF}{AB} = \frac{EH}{AD}$$

$$\frac{EF}{6.5} = \frac{6}{7.8}$$

$$EF = \frac{6 \times 6.5}{7.8} = 5 \text{ cm}$$

Similar shapes checklist

Corresponding angles equal. ✓

Corresponding sides in same ratio. ✓

Spotting similar triangles

Here are some similar triangles which often appear in questions.

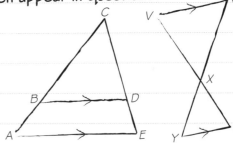

Triangle *ACE* is similar to triangle *BCD*.

Triangle *VWX* is similar to triangle *ZYX*.

Use the fact that *AB* is parallel to *PQ* to see which angles are equal.

The sides opposite the equal angles are corresponding.

Now try this

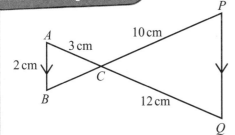

Diagram **NOT** accurately drawn

ACQ and *BCP* are straight lines. *AB* is parallel to *PQ*.

(a) Work out the length of *PQ*.

(2 marks)

(b) Work out the length of *BP*.

(3 marks)

grade **A**

 edexcel

Had a look ☐ Nearly there ☐ Nailed it! ☐

A*
A
B
C
D

Similar shapes 2

The relationship between similar shapes is defined by a SCALE FACTOR.
A and B are similar shapes. B is an enlargement of A with scale factor k.

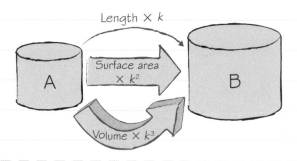

Length × k

Surface area × k^2

Volume × k^3

When a shape is enlarged by a linear scale factor k
- Enlarged surface area
  $= k^2 \times$ original surface area
- Enlarged volume
  $= k^3 \times$ original volume
- Enlarged mass $= k^3 \times$ original mass

Worked example

grade A*

A and **B** are two solid shapes which are mathematically similar.
The shapes are made from the same material.

A

B

The surface area of **A** is 50 cm².
The surface area of **B** is 18 cm².
The mass of **A** is 500 grams.
Calculate the mass of **B**.

$18 = k^2 \times 50$

$k^2 = \dfrac{18}{50} = 0.36$

$k = \sqrt{0.36} = 0.6$

Mass of **B** $= k^3 \times 500$
$\qquad\qquad = 0.6^3 \times 500 = 108$

So mass of **B** is 108 g.

EXAM ALERT!

Out of every 30 students who took the exam, only one got full marks on this question. To be that one you need to follow these steps.

1. Use the two surface areas given to write down a relationship involving k^2.

2. Solve the equation to find the value of k.

3. Multiply the mass of **A** by k^3 to find the mass of **B**.

Check it!
Shape **B** is smaller than shape **A** so it should have a smaller mass. ✓

This was a real exam question that caught students out – **be prepared!**   Results Plus

Comparing volumes

You can use k^3 to compare volume, mass or capacity.

$k = \dfrac{32}{16} = 2$

Volume of large bottle
$= 1.2 \times k^3$
$= 1.2 \times 8$
$= 9.6$ litres

1.2 litres   ←16 cm→   ←32 cm→

Now try this

grade A*

1. **X** and **Y** are two geometrically similar solid shapes.

   The total surface area of shape **X** is 450 cm².
   The total surface area of shape **Y** is 800 cm².
   The volume of shape **X** is 1350 cm³.
   Calculate the volume of shape **Y**.   **(3 marks)**

2. Two prisms, **A** and **B**, are mathematically similar.
   The volume of prism **A** is 12 000 cm³.
   The volume of prism **B** is 49 152 cm³.
   The total surface area of prism **B** is 9728 cm².
   Calculate the total surface area of prism **A**.
   (4 marks)

grade A*

edexcel

Bearings

A*
A
B
C
D

Bearings are measured CLOCKWISE from NORTH.

Bearings always have 3 FIGURES. You might need to add zeros if the angle is less than 100°. In this diagram the bearing of B from A is 048°.

You can measure a bearing bigger than 180° by measuring this angle and subtracting it from 360°.

The bearing of C from A is
$360° - 109° = 251°$

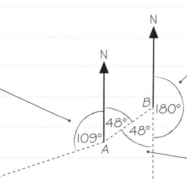

You can work out a reverse bearing by adding or subtracting 180°.

The bearing of A from B is
$180° + 48° = 228°$

These are alternate angles.

Worked example

grade **C**

The diagram shows the position of three buildings in a town. The bearing of the Hospital from the Art gallery is 072°. Work out the bearing of the Cinema from the Art gallery.

Diagram **NOT** accurately drawn

$72° + 180° = 252°$

$162 + 2x = 180$

$2x = 18$

$x = 9$

$72 + 9 = 81$

The bearing of the Cinema from the Art gallery is 081°.

1. Find the bearing of the Art gallery **from** the Hospital by adding 180°. Show this angle on the diagram.

2. Sketch the isosceles triangle. The size of the large angle is $252° - 90° = 162°$. Work out the size of each base angle.

3. Add one of the base angles to 72° to find the bearing of the Cinema from the Art gallery. Remember to use three figures in your bearing.

Now try this

You need to measure each bearing with a **protractor** and draw with a **ruler** and a **sharp pencil.**

edexcel

The diagram shows the positions of two ships, A and B.

A ship C is on a bearing of 064° from ship A. Ship C is also on a bearing of 290° from ship B.

(a) Draw an accurate diagram to show the position of ship C. Mark the position of ship C with a cross **✗**. Label it C. **(3 marks)**

grade **D**

Another ship D is on a bearing of 128° from ship E.

(b) Work out the bearing of ship E from ship D. **(2 marks)**

A*
A
B
C
D

Scale drawings and maps

This is a SCALE DRAWING of the
Queen Mary II cruise ship.

Scale = 1 : 1000

← 34.5 cm →

You can use the scale to work out the
length of the actual ship.

34.5 × 1000 = 34 500

The ship is 34 500 cm or 345 m long.

Map scales

Map scales can be
written in different
ways:

* 1 to 25 000
* 1 cm represents 25 000 cm
* 1 cm represents 250 m
* 4 cm represent 1 km

Worked example

grade D

The diagram shows a scale drawing
of a port and a lighthouse.

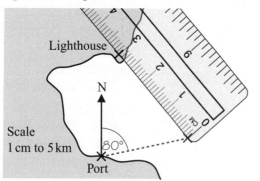

Lighthouse

N

Scale
1 cm to 5 km

80°

Port

A boat sails 12 km in a straight line on a bearing
of 080°.

(a) Mark the new position of the boat with a
cross.

(b) How far away is the boat from the
lighthouse? Give your answer in km.

15 km

For a reminder about bearings have a
look at page 69.

(a) Start by working out how far the
boat is from the port on the scale
drawing.

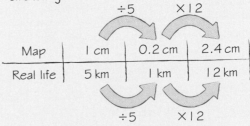

÷5   ×12

Map	1 cm	0.2 cm	2.4 cm
Real life	5 km	1 km	12 km

÷5   ×12

Now place the centre of your
protractor on the port with the zero
line pointing North. Put a dot at 80°.
Line up your ruler between the port
and the dot. Draw a cross 2.4 cm
from the port.

(b) Use a ruler to measure the distance
from the lighthouse to the boat.
3 cm on the drawing represents
15 km in real life.

Now try this

edexcel

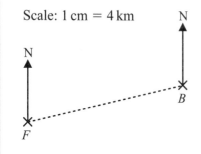

Scale: 1 cm = 4 km

N

N

×
B

×
F

The diagram shows the position of a farm F and a bridge B on a map.

(a) Measure and write down the bearing of B from F.

(1 mark)

grade D

A church C is on a bearing of 155° from the bridge B.
The church is 20 km from B.

(b) Mark the church with a cross (✗) and label it C.   **(2 marks)**

(c) How far away is church C from farm F?
Give your answer in km.

(2 marks)

Constructions

'Construct' means 'draw accurately using a ruler and compasses'. You should make sure you have a good pair of compasses with stiff arms and a sharp pencil in your exam.

Worked example grade D

1 Use a ruler and compasses to **construct** a triangle with sides of length 3 cm, 4 cm and 5.5 cm.

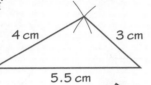

Draw and label one side with a ruler. Then use your compasses to find the other vertex.

Worked example grade C

3 Use a ruler and compasses to **construct** the perpendicular bisector of the line *AB*.

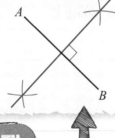

EXAM ALERT!

Use your compasses to draw intersecting arcs with centres *A* and *B*.

Less than half of students got full marks on this question.

You must show **all** your construction lines to give a full answer.

This was a real exam question that caught students out – **be prepared!**   ResultsPlus

Now try this   edexcel

Use a ruler and compasses to **construct** an angle of 30° at *P*.

You **must** show all your construction lines.   grade C

P _____ **(3 marks)**

Worked example grade C

2 Use a ruler and compasses to **construct** the bisector of angle *PQR*.

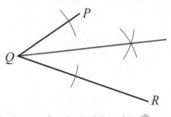

Mark points on each arm an equal distance from *Q*. Then use arcs to find a third point an equal distance from these two points.

Worked example grade C

4 Use a ruler and compasses to **construct** the perpendicular to the line segment *AB* that passes through point *P*.

Use your compasses to mark two points on the line an equal distance from *P*. Then widen your compasses and draw arcs with their centres at these two points.

Worked example grade C

5 Use a ruler and compasses to **construct** the perpendicular to the line segment *AB* that passes through point *P*.

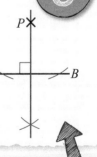

Use your compasses to mark two points on the line an equal distance from *P*. Keep the compasses the same and draw two arcs with their centres at these points.

-A*-
-A-
-B-
-C-
-D-

Loci

A LOCUS is a set of points which satisfy a condition. You can construct loci using a ruler and compasses. A set of points can lie inside a REGION rather than on a line or curve.

The locus of points which are 7 cm from A is the circle, centre A.

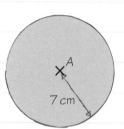

7 cm

The region of points less than 7 cm from A lies inside this circle.

The locus of points which are the SAME DISTANCE from B as from C is the perpendicular bisector of BC.

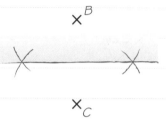

×B

×C

Points in the shaded region are closer to B than to C.

The locus of points which are 2 cm away from ST consists of two semicircles and two straight lines.

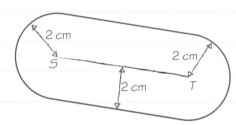

2 cm   2 cm
S
2 cm
T

Combining conditions

You can be asked to shade a region which satisfies more than one condition.

Here, the shaded region is more than 6 cm from point D and closer to line BC than to line AD.

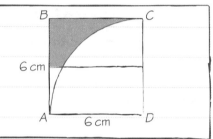

B   C
6 cm
A   6 cm   D

Worked example   grade C

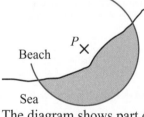

Beach   P×

Sea

The diagram shows part of a beach and the sea.
1 cm represents 20 m.

There is a lifeguard tower at point P.

Public swimming is allowed in a region of the sea less than 30 m from the lifeguard tower. Shade this region on the diagram.

Everything in red is part of the answer.

1 cm represents 20 m so 1.5 cm represents 30 m.

There is more about scale drawing on page 70.

Set your compasses to 1.5 cm. You can set your compasses accurately by placing the point **on top of** your ruler at the 0 mark.

Now try this   edexcel   grade C

The map shows part of a lake.

In a competition for radio-controlled boats, a competitor has to steer a boat so that its path between *AB* and *CD* is a straight line **and** this path is always the same distance from *A* as from *B*.

On the map, draw the path the boat should take.   **(2 marks)**

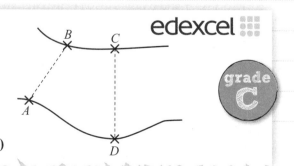

B   C

A

D

Translations, reflections and rotations

You might have to describe these transformations in your exam. To describe a translation you need to give a vector. To describe a reflection you need to give the equation of the mirror line. To describe a rotation you need to give the direction, the angle and the centre of rotation.

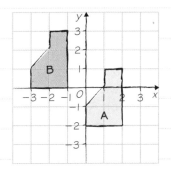

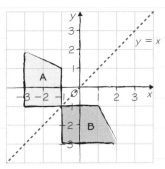

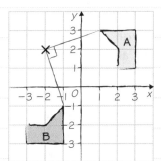

A to B: TRANSLATION by the vector $\begin{pmatrix} -3 \\ 2 \end{pmatrix}$.

A to B: REFLECTION in the line $y = x$.

A to B: ROTATION 90° clockwise about the point $(-2, 2)$.

You can ask for tracing paper in an exam. This makes it easy to rotate shapes and check your answers.

For all three transformations, lengths of sides do not change, angles in shapes do not change and B is congruent to A.

Worked example

grade C

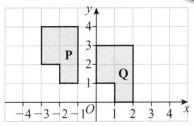

Describe <u>fully</u> the single transformation that will map shape **P** onto shape **Q**.

Translation by the vector $\begin{pmatrix} 3 \\ -1 \end{pmatrix}$

EXAM ALERT!

Only one-fifth of students got full marks for this.

First write the word 'translation'. Do **not** use words such as 'transformed' or 'moved'.

You must use a vector to describe the translation:

$$\begin{pmatrix} \text{movement to the right} \\ \text{movement up} \end{pmatrix}$$

Use negative numbers to describe movement to the left or down. Do **not** describe the movement as 'across 3 and down 1'.

This was a real exam question that caught students out – **be prepared!**
 Results**Plus**

Now try this

grade C　　edexcel ⠿

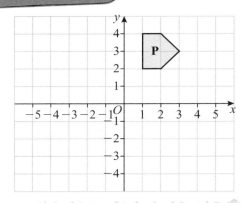

(a) On the grid, rotate the shaded shape **P** one quarter turn anticlockwise about *O*. Label the new shape **Q**.　　**(3 marks)**

(b) On the grid, translate the shaded shape **P** by the vector $\begin{pmatrix} -2 \\ -3 \end{pmatrix}$. Label the new shape **R**.　　**(2 marks)**

-A*-
-A-
-B-
-C-
-D-

Enlargements

To describe an enlargement you need to give the scale factor and the centre of enlargement.

The SCALE FACTOR of an enlargement tells you how much each length is multiplied by.

$$\text{Scale factor} = \frac{\text{enlarged length}}{\text{original length}}$$

Lines drawn through corresponding points on the object (A) and image (B) meet at the CENTRE OF ENLARGEMENT.

When the scale factor is between 0 and 1, image B is SMALLER than object A.

When the scale factor is negative, image B is on the OTHER SIDE of the centre of enlargement.

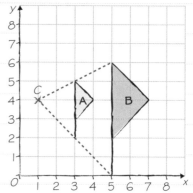

A to B: Each point on B is twice as far from C as the corresponding point on A.

Enlargement with scale factor 2, centre (1, 4).

For enlargements, angles in shapes do not change but lengths of sides do change.

Worked example

grade A

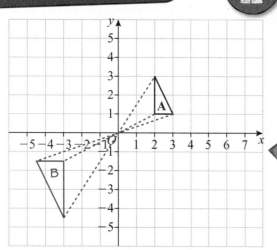

Enlarge triangle **A** by scale factor $-1\frac{1}{2}$ with centre of enlargement O.

1. Draw lines from each vertex through the centre of enlargement.
2. The scale factor is negative so the enlargement (**B**) is on the opposite side of the centre of enlargement and is upside down.
3. You can measure the distance from each point on **A** to the centre of enlargement. The corresponding point on **B** is $1\frac{1}{2}$ times the distance from the centre of enlargement.

Check it!
Each length on **B** should be $1\frac{1}{2}$ times the corresponding length on **A**. ✓
If you were given this completed diagram, could you describe the transformation? If you're going for an A*, you need to be comfortable with negative fractional scale factors.

Now try this

edexcel

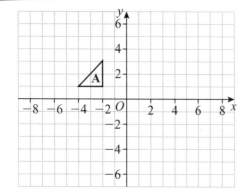

(a) Enlarge shape **A** by a scale factor $\frac{1}{2}$, centre (6, 1). Label the new shape **B**.   **(3 marks)**

(b) Enlarge shape **A** by a scale factor -2, centre O. Label the new shape **C**.   **(3 marks)**

grade A

Combining transformations

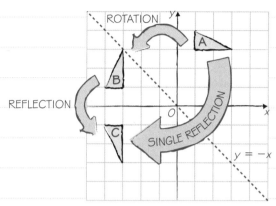

-A*-
-A-
-B-
-C-
-D-

You can describe two or more transformations using a single transformation.

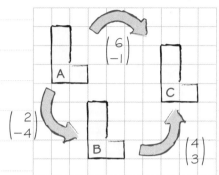

A to B to C: A translation $\begin{pmatrix} 2 \\ -4 \end{pmatrix}$ followed by a translation $\begin{pmatrix} 4 \\ 3 \end{pmatrix}$ is the same as a single translation $\begin{pmatrix} 6 \\ -1 \end{pmatrix}$.

A to B to C: A rotation 90° anticlockwise about O followed by a reflection in the x-axis is the same as a single reflection in the line $y = -x$.

Worked example

grade B

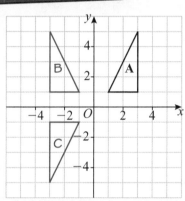

Triangle **A** is reflected in the y-axis to give **B**.

Triangle **B** is then reflected in the x-axis to give **C**.

Describe fully the **single** transformation that takes triangle **A** to triangle **C**.

Rotation 180° about the point O.

You need to draw both transformations on the diagram. Remember that triangle **C** is a reflection of triangle **B**.

To describe a rotation fully you need to write 'rotation' and give the angle and centre of rotation. For a rotation of 180° you don't need to give a direction.

Describe fully...

A translation: vector of translation. ✓

A reflection: equation of mirror line. ✓

A rotation: angle of turn, direction of turn and centre of rotation. ✓

An enlargement: scale factor and centre of enlargement. ✓

Now try this

 edexcel

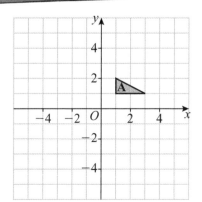

Triangle **A** is enlarged by a scale factor of -1, centre O to give triangle **B**.

Triangle **B** is translated by the vector $\begin{pmatrix} 4 \\ -2 \end{pmatrix}$ to give triangle **C**.

Describe fully the single transformation that will map triangle **A** onto triangle **C**.   **(4 marks)**

grade A*

A*
A
B
C
D

Line segments

The section of a straight line between two points, P and Q, is called a line segment.

For a reminder of how to find the midpoint of a line segment have a look at page 26.

You can use Pythagoras' theorem to find the length of the line segment PQ.

Start by drawing a right-angled triangle with PQ as its hypotenuse.

For a reminder about Pythagoras' theorem have a look at page 60.

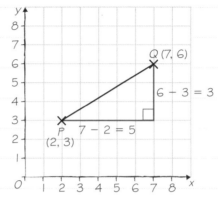

$PQ^2 = 5^2 + 3^2$
$= 25 + 9$
$= 34$
$PQ = \sqrt{34}$
$= 5.8309...$
$= 5.83$ (to 3 s.f.)

Worked example

grade C

Point A has coordinates $(2, 5)$.
Point B has coordinates $(3, -2)$.
Calculate the length of the line segment AB.
Give your answer correct to 3 significant figures.

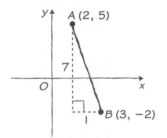

$AB^2 = 1^2 + 7^2$
$= 50$
$AB = \sqrt{50}$
$= 7.07106... = 7.07$ (to 3 s.f.)

1. Sketch x- and y-axes.
2. Mark points A and B on your sketch.
3. Draw a right-angled triangle with AB as its hypotenuse.
4. Work out the lengths of the two short sides of the triangle.
5. Use Pythagoras' theorem to work out the length of AB.
6. Round your answer to 3 significant figures.

Using a formula

If P has coordinates (x_1, y_1) and Q has coordinates (x_2, y_2) then the length of the line segment PQ is

$$\sqrt{(x_2 - x_1)^2 + (y_2 - y_1)^2}$$

Surds

You might be asked to give an EXACT ANSWER to a calculation, or to write your answer in a certain FORM.

You can write $\sqrt{50}$ in the form $k\sqrt{2}$ by finding a factor which is a square number.

$\sqrt{50} = \sqrt{25 \times 2} = \sqrt{25} \times \sqrt{2} = 5\sqrt{2}$ k has to be an integer.

For a reminder about writing answers as surds have a look at page 16.

Now try this

grade C

edexcel

P is the point with coordinates $(5, -1)$.
Q is the point with coordinates $(-3, -5)$.
Calculate the exact length of PQ.   **(4 marks)**

Remember to draw a sketch to help you. The question asks for the 'exact' length, so give your answer in surd form.

Trigonometry 1

You can use the trigonometric ratios to find the size of an angle in a right-angled triangle. You need to know the lengths of two sides of the triangle.

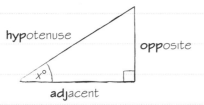

The sides of the triangle are labelled relative to the ANGLE you need to find.

Trigonometric ratios

$\sin x° = \dfrac{\text{opp}}{\text{hyp}}$ (remember this as S^O_H)

$\cos x° = \dfrac{\text{adj}}{\text{hyp}}$ (remember this as C^A_H)

$\tan x° = \dfrac{\text{opp}}{\text{adj}}$ (remember this as T^O_A)

You can use $S^O_H C^A_H T^O_A$ to remember these rules for trig ratios.
These rules only work for RIGHT-ANGLED triangles.

Worked example

grade **B**

Calculate the size of angle a in this right-angled triangle.

Give your answer correct to 3 significant figures.

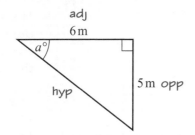

$\tan a° = \dfrac{\text{opp}}{\text{adj}} = \dfrac{5}{6}$

$a° = 39.805\,571\,09 = 39.8°$ (to 3 s.f.)

Label the **hyp**otenuse first — it's the longest side.

Then label the side **adj**acent to the angle you want to work out.

Finally label the side **opp**osite the angle you want to work out.

Remember $S^O_H C^A_H T^O_A$. You know **opp** and **adj** here so use T^O_A.

Do **not** 'divide by tan' to get a on its own. You need to use the $\tan^{-1}$ function on your calculator.

$\tan^{-1}\left(\dfrac{5}{6}\right)$
     39.80557109

Write down all the figures on your calculator display then round your answer.

Using your calculator

To find a missing angle using trigonometry you have to use one of these functions.

$$\sin^{-1} \quad \cos^{-1} \quad \tan^{-1}$$

These are called INVERSE TRIGONOMETRIC functions. They are the inverse operations of sin, cos and tan.

Make sure that your calculator is in degree mode. Look for the **D** symbol at the top of the display.

Now try this

Work out the value of x.
Give your answer correct to 1 decimal place.
(3 marks)

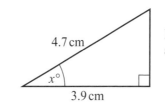

Diagram **NOT** accurately drawn

grade **B**

edexcel

-A*-
-A-
-B-
-C-
-D-

Trigonometry 2

You can use the trigonometric ratios to find the length of a missing side in a right-angled triangle. You need to know the length of another side and the size of one of the acute angles.

Worked example

Calculate the length of the side marked a.
Give your answer correct to 3 significant figures.

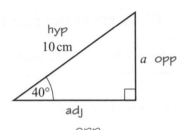

Diagram **NOT**
accurately drawn

hyp
10 cm

a opp

40°

adj

$S^O_H \checkmark\ C^A_H \checkmark\ T^O_A \checkmark$

$$\sin x° = \frac{\text{opp}}{\text{hyp}}$$

$$\sin 40° = \frac{a}{10}$$

$$a = 10 \times \sin 40°$$
$$= 6.42787...$$
$$= 6.43\,\text{cm (to 3 s.f.)}$$

Label the sides of the triangle relative to the 40° angle. Write $S^O_H C^A_H T^O_A$ and tick the pieces of information you have. You need to use S^O_H here.

Write the values you know in the rule and replace **opp** with a. You can solve this equation to find the value of a.

Write down at least four figures of the calculator display before giving your final answer.

Check it!
Side a must be shorter than the hypotenuse. 6.43 cm looks about right. ✓

Angles of elevation and depression

Some trigonometry questions will involve angles of elevation and depression.

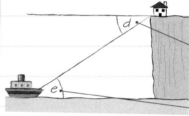

The angle of depression of the ship from the house.

The angle of elevation of the house from the ship.

Angles of elevation and depression are always measured from the horizontal.

In this diagram, $d = e$ because they are alternate angles.

Now try this

grade B

grade A

edexcel

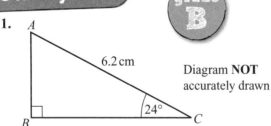

1.

A

6.2 cm

Diagram **NOT**
accurately drawn

B

24°

C

Angle $ABC = 90°$.
Angle $ACB = 24°$.
$AC = 6.2$ cm.
Calculate the length of BC.
Give your answer correct to 3 significant figures.   **(3 marks)**

2.

B

57°

7 cm

A 65°

D

C

Diagram **NOT**
accurately drawn

ADC is a straight line with BD perpendicular to AC.
$AB = 7$ cm.
Angle $BAD = 65°$.
Angle $CBD = 57°$.
Calculate the length of AC.
Give your answer correct to 3 significant figures.   **(6 marks)**

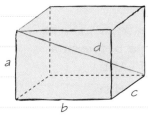

Pythagoras in 3-D

If you're going for a very top grade you need to be able to use Pythagoras' theorem in 3-D shapes.

You can use Pythagoras' theorem to find the length of the longest diagonal in a cuboid.

You can also use Pythagoras to find missing lengths in pyramids and cones.

For a reminder about Pythagoras' theorem have a look at page 60.

$$a^2 + b^2 + c^2 = d^2$$

Why does it work?

You can use 2-D Pythagoras twice to show why the formula for 3-D Pythagoras works.

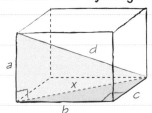

$$x^2 = b^2 + c^2$$

$$d^2 = a^2 + x^2$$
$$= a^2 + b^2 + c^2$$

Worked example

grade A

The diagram shows a cuboid. Work out the length of PQ.

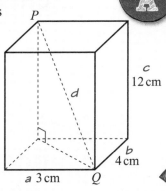

$$d^2 = a^2 + b^2 + c^2$$
$$PQ^2 = 3^2 + 4^2 + 12^2$$
$$= 169$$
$$PQ = \sqrt{169} = 13$$
So PQ is 13 cm.

Write out the formula for Pythagoras in 3-D. Label the sides of the cuboid a, b and c, and label the long diagonal d. You could also answer this question by sketching two right-angled triangles and using 2-D Pythagoras.

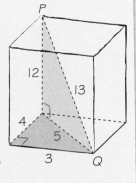

Check it!
The diagonal must be longer than any of the other three lengths.
13 cm looks about right. ✓

Now try this

grade A

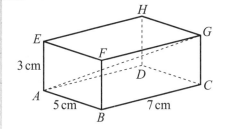

Diagram **NOT** accurately drawn

edexcel

The diagram represents a cuboid $ABCDEFGH$.
$AB = 5$ cm
$BC = 7$ cm
$AE = 3$ cm
Calculate the length of AG.
Give your answer correct to 3 significant figures.
(2 marks)

A*
A
B
C
D

Trigonometry in 3-D

You can use $S^O_H C^A_H T^O_A$ to find the angle between a LINE and a PLANE.

You might need to combine trigonometry and Pythagoras' theorem when you are solving 3-D problems.

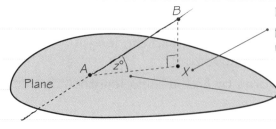

Point X is directly below point B, so ABX is a right-angled triangle.

Angle z is the angle between the line and the plane.

Plane

Worked example

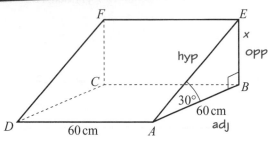

The diagram shows a triangular prism.
Work out the angle that the line DE makes with the plane $ABCD$.

$\triangle ABE$: you know one angle and the adjacent side. You are looking for the opposite side, so use T^O_A.

$\triangle DAB$: you know two sides so you can use Pythagoras' theorem.

$\triangle DEB$: you know the opposite and adjacent sides so use T^O_A. Use $\tan^{-1}$ to find the value of z.

Do **not** round any of your answers until the end — write down at least six figures from each calculator display.

$\tan 30° = \dfrac{x}{60}$

$x = 60 \times \tan 30°$

$\quad = 34.6410...\,cm$

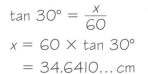

$y^2 = 60^2 + 60^2$

$\quad = 7200$

$y = \sqrt{7200} = 84.8528...\,cm$

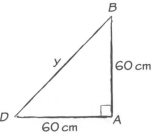

$\tan z° = \dfrac{34.6410...}{84.8528...} = 0.408\,24...$

$z° = \tan^{-1} 0.408\,24... = 22.2076...$

$\quad = 22.2°$ (to 3 s.f.)

Now try this

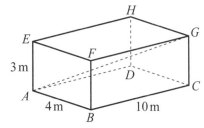

Diagram **NOT** accurately drawn

The diagram represents a cuboid $ABCDEFGH$.
(a) Calculate the length of AG.
   Give your answer correct to 3 significant figures.

(2 marks)

(b) Calculate the size of the angle between AG and the face $ABCD$.
   Give your answer correct to 1 decimal place.

(2 marks)

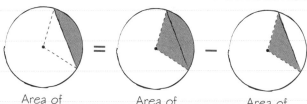

Triangles and segments

A*
A
B
C
D

When you know the lengths of two sides and the angle BETWEEN THEM, the area of any triangle can be found using this formula.

$$\text{Area} = \tfrac{1}{2} ab \sin C$$

You can use this formula for ANY triangle. You don't need to have a right angle.

This formula is on the formula sheet.

Areas of segments

A chord divides a circle into two SEGMENTS.

Area of segment = Area of whole sector − Area of triangle

For a reminder about areas of sectors see page 58.

Worked example

grade A*

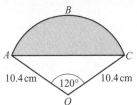

The diagram shows a sector of a circle with centre O.

Calculate the area of the shaded segment ABC.
Give your answer correct to 3 significant figures.

Whole sector $OABC$:
$$\text{Area} = \frac{120}{360} \times \pi \times 10.4^2$$
$$= 113.2648\ldots \text{ cm}^2$$

Triangle OAC:
$$\text{Area} = \tfrac{1}{2} \times 10.4 \times 10.4 \times \sin 120°$$
$$= 46.8346\ldots \text{ cm}^2$$

Shaded segment ABC:
$$\text{Area} = 113.2648\ldots - 46.8346\ldots$$
$$= 66.4302\ldots$$
$$= 66.4 \text{ cm}^2 \text{ (to 3 s.f)}$$

This is a very common exam question! You need to be able to calculate the area of a sector and a triangle.

To give a full answer you need to keep track of your working. Make sure you write down exactly what you are calculating at each step. Remember that 10.4 cm is the length of one side of the triangle **and** the radius of the circle. Make sure you don't round too soon. Write down all the figures from your calculator display at each step. Only round your **final answer** to 3 significant figures.

Which formula?

If you know the base and the vertical height:
$$\text{Area} = \tfrac{1}{2} \times \text{base} \times \text{vertical height}$$
$$= \tfrac{1}{2} \times 6 \times 2.1$$
$$= 6.3 \text{ cm}^2$$

If you know two sides and the included angle:
$$\text{Area} = \tfrac{1}{2} ab \sin C$$
$$= \tfrac{1}{2} \times 5 \times 6 \times \sin 25°$$
$$= 6.3\ldots \text{ cm}^2$$

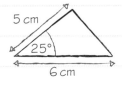

Now try this

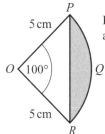

Diagram **NOT** accurately drawn

edexcel

The diagram shows a sector $OPQR$ of a circle with centre O.

$OP = OR = 5$ cm
Angle $POR = 100°$

Calculate the area of the shaded segment PQR.

Give your answer correct to 3 significant figures.

grade A*

(4 marks)

-A*-
-A-
-B-
-C-
-D-

The sine rule

The SINE RULE applies to any triangle. You don't need a right angle. You label the angles of the triangle with capital letters and the sides with lower case letters. Each side has the same letter as its OPPOSITE angle.

$$\frac{a}{\sin A} = \frac{b}{\sin B} = \frac{c}{\sin C}$$

This version is given on the formula sheet. Use it to find a missing side.

$$\frac{\sin A}{a} = \frac{\sin B}{b} = \frac{\sin C}{c}$$

Learn this version. It's useful for finding a missing angle.

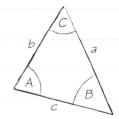

Worked example   grade A

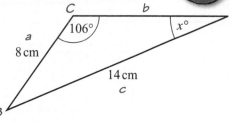

Work out the size of the angle marked x.

$$\frac{\sin A}{a} = \frac{\sin C}{c}$$

$$\frac{\sin x^\circ}{8} = \frac{\sin 106^\circ}{14}$$

$$\sin x^\circ = \frac{8 \times \sin 106^\circ}{14}$$

$$= 0.549\,292\,3977$$

$$x^\circ = 33.318\,48\ldots^\circ = 33.3^\circ \text{ (to 3 s.f.)}$$

Golden rule

To use the sine rule, you need to know either two angles and a side (ASA) or two sides and a non-included angle (SSA).

This is not a right-angled triangle so you can't use $S^O_HC^A_HT^O_A$. You know two sides and a non-included angle (SSA) so you can use the sine rule.

You need to find an angle so use the 'upside down' version of the sine rule:

$$\frac{\sin A}{a} = \frac{\sin B}{b} = \frac{\sin C}{c}$$

You are not interested in side b or angle B so ignore this part of the rule.

Substitute in the values you know and solve the equation to find x. Use the $\sin^{-1}$ function on your calculator.

Worked example   grade A

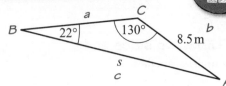

Work out the length of the side marked s.

$$\frac{c}{\sin C} = \frac{b}{\sin B}$$

$$\frac{s}{\sin 130^\circ} = \frac{8.5}{\sin 22^\circ}$$

$$s = \frac{8.5 \times \sin 130^\circ}{\sin 22^\circ}$$

$$= 17.381\,909\ldots$$

The length is 17.4 m (to 3 s.f.).

You know two angles and a side (ASA) so you can use the sine rule.

Check it!
The greater side length is opposite the greater angle. ✓

Now try this   edexcel   grade A

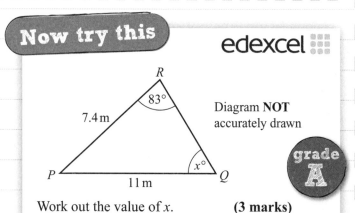

Diagram **NOT** accurately drawn

Work out the value of x.   **(3 marks)**

The cosine rule

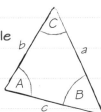

The COSINE RULE applies to any triangle. You don't need a right angle.

You usually use the cosine rule when you are given two sides and the included angle (SAS) or when you are given three sides and want to work out an angle (SSS).

$$a^2 = b^2 + c^2 - 2bc \cos A$$

This version is on the formula sheet. Use it to find a missing side.

$$\cos A = \frac{b^2 + c^2 - a^2}{2bc}$$

Learn this version. It's useful for finding a missing angle.

Worked example   grade A

Calculate the length of PQ.

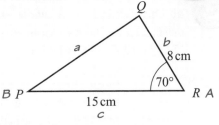

$a^2 = b^2 + c^2 - 2bc \cos A$
$PQ^2 = 8^2 + 15^2 - 2 \times 8 \times 15 \times \cos 70°$
$\quad = 206.9151...$
$PQ = 14.4 \text{ cm (to 3 s.f.)}$

This is not a right-angled triangle so you can't use $S^O{}_HC^A{}_HT^O{}_A$. You know two sides and the included angle (SAS) so you can use the cosine rule.

Let a represent the side you want to calculate.

Substitute the values you are given into the formula.

Use BIDMAS when you work out the value of the right-hand side. Do **not** work out $8^2 + 15^2 - 2 \times 8 \times 15$ and then multiply by $\cos 70°$.

Worked example   grade A

Calculate the size of the angle marked x.

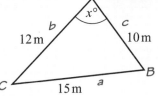

$\cos A = \dfrac{b^2 + c^2 - a^2}{2bc}$

$\cos x° = \dfrac{12^2 + 10^2 - 15^2}{2 \times 12 \times 10}$

$\quad = \dfrac{19}{240}$

$x° = 85.4593...° = 85.5°$ (to 3 s.f.)

You know all three sides (SSS) so you can use the cosine rule.

You need to find an angle so use this version:

$$\cos A = \frac{b^2 + c^2 - a^2}{2bc}$$

Let A represent the angle you want to calculate.

Use the **cos⁻¹** function on your calculator to find the angle.

Sine or cosine?

Use the sine rule when a problem involves two sides and two angles.

Use the cosine rule when a problem involves three sides and one angle.

Look at page 89 for an A* cosine rule question.

Now try this   edexcel

Calculate the length of AB.
Give your answer correct to 3 significant figures.

(3 marks)

grade A

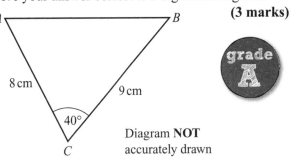

Diagram **NOT** accurately drawn

Circle facts

You need to know the names of the different parts of a circle.

Diameter = radius × 2

Tangent

The other parts of a circle are shown on pages 57 and 81.

When you are solving circle problems:

- correctly identify the angle to be found
- use all the information given in the question
- mark all calculated angles on the diagram
- give a reason for each step of your working.

You might need to use angle facts about triangles, quadrilaterals and parallel lines in circle questions. There is a list of angle facts on page 52.

Key circle facts

1 The angle between a radius and a tangent is 90°.

2 Two tangents which meet at a point outside a circle are the same length.

3 A triangle which has one vertex at the centre of a circle and two vertices on the circumference is an ISOSCELES TRIANGLE.

Each short side of the triangle is a radius, so they are the same length.

Remember that the base angles of an isosceles triangle are equal.

Worked example

grade **B**

A and B are points on the circumference of a circle centre O. AC and BC are both tangents to the circle. Angle $BCA = 42°$. Work out the size of the angle marked x.

$AC = BC$ (tangents from a point outside a circle are the same length)

$\angle ABC = \dfrac{180° - 42°}{2} = 69°$

(base angles in an isosceles $\triangle$ are equal, and angles in a $\triangle$ add up to 180°)

$x + 69° = 90°$ (angle between a tangent and a radius = 90°)

$x = 21°$

$AC = BC$, so mark these lines with a dash. Make sure you write down the circle fact you are using as well. To give a full answer, you have to give a reason for each step of your working.

Now try this

Angle LMN is the angle at M with 'arms' ML and MN.

edexcel

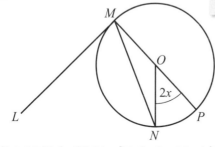

M, N and P are points on a circle, centre O.
LM is a tangent to the circle.
MOP is a straight line.
Angle $NOP = 2x°$.
Prove that angle $LMN = (90 - x)°$ **(4 marks)**

grade **A**

Circle theorems

1 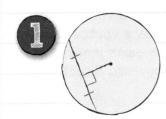 The perpendicular from a chord to the centre of the circle bisects the chord.

2 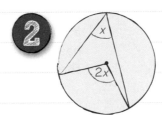 The angle at the centre of the circle is twice the angle on the circumference.

3  The angle in a semicircle is 90°.

4  Angles in the same segment are equal.

5  Opposite angles of a cyclic quadrilateral add up to 180°.

6 The angle between a tangent and a chord is equal to the angle in the alternate segment.
This is called the ALTERNATE SEGMENT THEOREM.

See page 84 for more circle facts.

Worked example

grade **A***

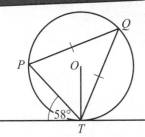

Diagram **NOT** accurately drawn

P, Q and T are points on the circumference of a circle, centre O.
The line ATB is the tangent at T to the circle.
$PQ = TQ$
Angle $ATP = 58°$
Calculate the size of angle OTQ.
Give a reason for each stage in your working.

$\angle ATO = 90°$
(angle between a radius and a tangent)
So $\angle PTO = 90° - 58° = 32°$
$\angle PQT = 58°$
(alternate segment theorem)
$\angle QPT = \angle QTP$
(base angles of an isosceles triangle)
So $\angle QTP = \dfrac{180° - 58°}{2} = 61°$
So $\angle OTQ = 61° - 32° = 29°$

EXAM ALERT!

About 80% of students got 0 marks on this question.
Look for angles you can work out and write them on the diagram.
Give reasons for **each step** of your working.
If a diagram has a triangle and a tangent in it, see if you can use the alternate segment theorem.

This was a real exam question that caught students out – **be prepared!**   Results**Plus**

Now try this

grade **A**

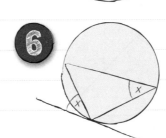

edexcel

Diagram **NOT** accurately drawn

(a) (i) Work out the value of x.   **(1 mark)**
     (ii) Give a reason for your answer. **(1 mark)**
(b) (i) Work out the value of y.   **(1 mark)**
     (ii) Give a reason for your answer. **(1 mark)**

Vectors

A vector has a MAGNITUDE (or size) and a DIRECTION.

This vector can be written as **a**, $\overrightarrow{AB}$ or $\begin{pmatrix} 2 \\ 5 \end{pmatrix}$.

You can multiply a vector by a number. The new vector has a different length but the same direction.

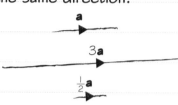

If **b** is a vector then −**b** is a vector with the same length but opposite direction.

Worked example

grade A*

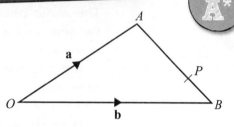

OAB is a triangle.
$\overrightarrow{OA} = \mathbf{a}$, $\overrightarrow{OB} = \mathbf{b}$

(a) Find the vector $\overrightarrow{AB}$ in terms of **a** and **b**.

$$\overrightarrow{AB} = \overrightarrow{AO} + \overrightarrow{OB}$$
$$= -\mathbf{a} + \mathbf{b}$$

P is the point on AB so that $AP : PB = 2 : 1$

(b) Find the vector $\overrightarrow{OP}$ in terms of **a** and **b**. Give your answer in its simplest form.

$$\overrightarrow{AP} = \tfrac{2}{3}\overrightarrow{AB}$$
$$= \tfrac{2}{3}(-\mathbf{a} + \mathbf{b})$$
$$\overrightarrow{OP} = \overrightarrow{OA} + \overrightarrow{AP}$$
$$= \mathbf{a} + \tfrac{2}{3}(-\mathbf{a} + \mathbf{b})$$
$$= \mathbf{a} - \tfrac{2}{3}\mathbf{a} + \tfrac{2}{3}\mathbf{b}$$
$$= \tfrac{1}{3}\mathbf{a} + \tfrac{2}{3}\mathbf{b}$$

Adding vectors

You can add vectors using the TRIANGLE LAW. You trace a path along the added vectors to find the new vector.

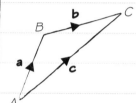

$\mathbf{a} + \mathbf{b} = \mathbf{c}$

c is the resultant vector of **a** and **b**.

$A \rightarrow B \rightarrow C$ is the same as $A \rightarrow C$

EXAM ALERT!

Nearly three-quarters of students got no marks on this question.

$AP : PB = 2 : 1$. There are $2 + 1 = 3$ parts in this ratio. So P is $\tfrac{2}{3}$ of the way along AB. This means that $\overrightarrow{AP} = \tfrac{2}{3}\overrightarrow{AB}$. Use the expression for $\overrightarrow{AB}$ from part (a) to write $\overrightarrow{AP}$ in terms of **a** and **b**.

Be careful: $\tfrac{2}{3}\overrightarrow{AB} \neq \tfrac{2}{3} - \mathbf{a} + \mathbf{b}$.

To write your expression in its simplest form, multiply out the brackets and collect any like terms.

This was a real exam question that caught students out – **be prepared!**

Now try this

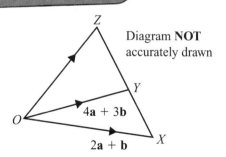

Diagram **NOT** accurately drawn

edexcel

XYZ is a straight line.
$XY : YZ = 2 : 3$

(a) Express the vector $\overrightarrow{XY}$ in terms of **a** and **b**. Give your answer in its simplest form. **(2 marks)**

(b) Express the vector $\overrightarrow{OZ}$ in terms of **a** and **b**. Give your answer in its simplest form. **(3 marks)**

grade A*

Solving vector problems

Parallel vectors

If one vector can be written as a MULTIPLE of the other then the vectors are PARALLEL.

In this parallelogram M is the midpoint of DC.
AB is parallel to DM so $\overrightarrow{DM} = \frac{1}{2}\overrightarrow{AB}$.

Remember that AB means the line segment AB (or the length of the line segment AB). $\overrightarrow{AB}$ means the vector which takes you from A to B.

The magnitude of a vector

You can use Pythagoras' theorem to calculate the magnitude (size) of a vector.

$$\mathbf{a} = \begin{pmatrix} 12 \\ 5 \end{pmatrix}$$

Magnitude of $\mathbf{a} = \sqrt{5^2 + 12^2} = 13$

Worked example

grade A*

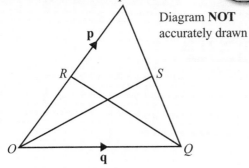

Diagram **NOT** accurately drawn

R is the midpoint of OP.
S is the midpoint of PQ.

(a) Find $\overrightarrow{OS}$ in terms of $\mathbf{p}$ and $\mathbf{q}$.

$\overrightarrow{PQ} = \overrightarrow{PO} + \overrightarrow{OQ} = -\mathbf{p} + \mathbf{q}$
$\overrightarrow{PS} = \frac{1}{2}\overrightarrow{PQ} = \frac{1}{2}(-\mathbf{p} + \mathbf{q})$
$\overrightarrow{OS} = \overrightarrow{OP} + \overrightarrow{PS}$
$\quad = \mathbf{p} + \frac{1}{2}(-\mathbf{p} + \mathbf{q})$
$\quad = \mathbf{p} - \frac{1}{2}\mathbf{p} + \frac{1}{2}\mathbf{q}$
$\quad = \frac{1}{2}\mathbf{p} + \frac{1}{2}\mathbf{q}$

(b) Show that RS is parallel to OQ.

$\overrightarrow{RO} = -\frac{1}{2}\mathbf{p}$
$\overrightarrow{RS} = \overrightarrow{RO} + \overrightarrow{OS}$
$\quad = -\frac{1}{2}\mathbf{p} + \frac{1}{2}\mathbf{p} + \frac{1}{2}\mathbf{q}$
$\quad = \frac{1}{2}\mathbf{q}$
$\overrightarrow{RS}$ is a multiple of $\overrightarrow{OQ}$ so RS is parallel to OQ.

(a) S is the midpoint of PQ so you know that $\overrightarrow{PS} = \frac{1}{2}\overrightarrow{PQ}$. You can work out $\overrightarrow{PQ}$ in terms of $\mathbf{p}$ and $\mathbf{q}$ by tracing from $P \rightarrow O \rightarrow Q$. Now use the triangle law to write $\overrightarrow{OS}$ in terms of $\mathbf{p}$ and $\mathbf{q}$.

(b) You already know that $\overrightarrow{OQ} = \mathbf{q}$. You need to show that $\overrightarrow{RS}$ can be written as a multiple of $\mathbf{q}$.

R is the midpoint of OP.
So $\overrightarrow{OR} = \frac{1}{2}\mathbf{p}$ and $\overrightarrow{RO} = -\frac{1}{2}\mathbf{p}$.
Trace from $R \rightarrow O \rightarrow S$. You can use the expression for $\overrightarrow{OS}$ that you worked out in part (a).

Now try this

edexcel

grade A*

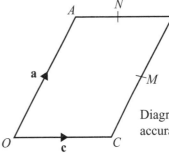

Diagram **NOT** accurately drawn

$OABC$ is a parallelogram.
M is the midpoint of CB.
N is the midpoint of AB.

(a) Find, in terms of $\mathbf{a}$ and/or $\mathbf{c}$, the vectors
   (i) $\overrightarrow{MB}$ (ii) $\overrightarrow{MN}$ **(2 marks)**
(b) Show that CA is parallel to MN.   **(2 marks)**

Had a look ☐ Nearly there ☐ Nailed it! ☐

Problem-solving practice

About half of the questions on your exam will need problem-solving skills.

These skills are sometimes called AO2 and AO3.

Practise using the questions on the next two pages.

For these questions you might need to:

• choose which mathematical technique or skill to use

• apply a technique in a new context

• plan your strategy to solve a longer problem

• show your working clearly and give reasons for your answers.

1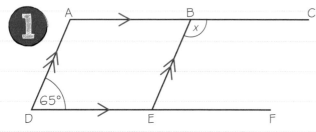

Diagram **NOT** accurately drawn

ABED is a parallelogram.

ABC and *DEF* are straight lines.

Find the size of the angle marked *x*.

You must give reasons to explain your

answer. (4 marks)

Solving angle problems p. 52

Just writing down the size of angle *x* is not enough. For **each step** of your working, write down any angle you have worked out **and** the angle fact or property you used.

grade **D**

TOP TIP

It's a good idea to write missing angles on the diagram as you work them out, but you still need to write down the **reason** for each step of your working.

---

2 *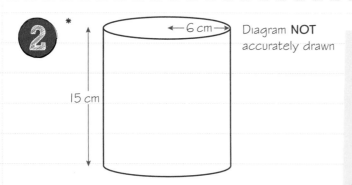

Diagram **NOT** accurately drawn

Jenny fills some empty flowerpots completely with compost.

Each flowerpot is in the shape of a cylinder of height 15 cm and radius 6 cm.

Jenny has a 15 litre bag of compost.

She fills up each flowerpot completely.

How many flowerpots can she fill completely?

You must show your working. (4 marks)

Circles and cylinders p. 57

You need to remember that 1 litre = 1000 cm³.

You are trying to work out how many flowerpots Jenny can fill **completely** so you'll need to round your final answer **down**.

grade **C**

TOP TIP

If a question has a * next to it, then there are marks available for QUALITY OF WRITTEN COMMUNICATION. This means you must show all your working and write your answer clearly with the correct units.

Problem-solving practice

 A ladder is 6 m long.

The ladder is placed on horizontal ground, resting against a vertical wall.

The instructions for using the ladder say that the bottom of the ladder must not be closer than 1.5 m to the bottom of the wall.

How far up the wall can the ladder reach if the instructions are followed?

Give your answer correct to 1 decimal place.   (3 marks)

Pythagoras' theorem p. 60

You should definitely draw a sketch to show the information in the question.

grade C

TOP TIP

Be careful when you are working out the length of a **short** side using Pythagoras' theorem.
Remember: $short^2 + short^2 = long^2$
$short^2 = long^2 - short^2$

---

4 Matt decides to make a bell. He mixes copper and tin to make the metal for the bell.

He has 270 kg of copper and $0.01 \, m^3$ of tin.

The density of copper is $9000 \, kg/m^3$.

The density of tin is $7300 \, kg/m^3$.

Work out the density of the bell.   (6 marks)

Density p. 65
Rearranging formulae p. 30

grade A

There's lots of work to do here, so plan your strategy before you start. You need to know the total mass and the total volume of the bell to calculate its density. So you need to work out the mass and volume of each type of metal using the formula **mass = volume × density** and then add them together.

TOP TIP

If mass is measured in kg and volume is measured in m^3, the units of density will be kg/m^3.

5

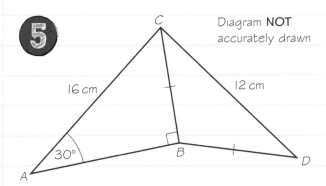

Diagram **NOT** accurately drawn

$AC = 16 \, cm$. $CD = 12 \, cm$.

$BC = BD$.

Angle $ABC = 90°$. Angle $CAB = 30°$.

Calculate the area of triangle BCD.

Give your answer correct to 3 significant figures.   (6 marks)

The cosine rule p. 83
Triangles and segments p. 81

grade A*

You need more information about triangle BCD before you can calculate its area.
Triangle ABC is right angled so you can use $S^O_H C^A_H T^O_A$ to work out the length of BC. Then use the cosine rule to work out the size of angle BCD. Finally, you can use the formula $A = \frac{1}{2}ab \sin C$ to work out the area of triangle BCD.

TOP TIP

Write down any values from your calculator to at least 4 decimal places before rounding your final answer to the required degree of accuracy.

Collecting data

You need to know what makes a question good or bad in a survey. Look at this example and then read the comments.

Internet use survey

✗ It's hard to know what this question means. Add at least 4 response boxes to improve the question:
☐ Laptop computer
☐ Desktop computer
☐ Smartphone
☐ Other

1. What do you use to access the internet?

   ....

2. How much time do you spend on the internet each day?

   ☐ Not very much   ☐ Average   ☐ A lot

✗ These responses could mean different things to different people. It would be better to ask how many hours they spend on the internet each day.

3. How many times a week do you check your email?

   ☐ 1–5   ☐ 5–10   ☐ 10–15   ☐ Every day

✗ The response boxes overlap. These response boxes would be better:
☐ 0–5   ☐ 6–10
☐ 11–15
☐ 16 or more

✗ People aren't very likely to answer this question truthfully. Don't ask people to reveal embarrassing or personal information.

4. Have you ever downloaded films illegally from the internet?

   ....

5. Do you agree that the BBC iPlayer is very easy to use?

   ☐ Yes   ☐ No

✗ This is a **biased** question. People are more likely to agree with you. A better question would be: The BBC iPlayer is easy to use.
☐ Agree   ☐ Disagree
☐ Neither

Worked example

grade D

Katya wants to find out information about the numbers of men, women, boys and girls using a sports hall.

Design a suitable <u>data collection sheet</u> to collect the information.

	Tally	Frequency
Men		
Women		
Boys		
Girls		

EXAM ALERT!

More than 50% of students got 0 marks for this question.

A data collection sheet (or table) must include a 'Tally' and a 'Frequency' column.

Do not design a questionnaire when asked for a data collection table.

This was a real exam question that caught students out – **be prepared!**   ResultsPlus

Now try this

For part (a), make sure you identify two completely **different** things.

edexcel

Toby wants to find out how many text messages people send.

He uses this question on a questionnaire.

How many text messages have you sent on your mobile phone?

☐ 0–10   ☐ 10–20   ☐ 20–30   ☐ 30 or more

(a) Write down two things wrong with this question.   **(2 marks)**   **grade D**

Toby also wants to find out how much time people spend talking on their mobile phones.

(b) Design a suitable question Toby could use for his questionnaire.
You must include some response boxes.   **(2 marks)**   **grade C**

Two-way tables

You can answer questions on two-way tables by adding or subtracting.

	Year 7	Year 8	Year 9	Total
Vegetarian	14	22	25	61
Not vegetarian	72	63	54	189
Total	86	85	79	250

There were 61 vegetarians in total.

In total 250 students were surveyed.

There were 86 Year 7 students surveyed.

There were 63 non-vegetarians in Year 8.

Worked example

The two-way table shows some information about the lunch arrangements of 85 students.

	School lunch	Packed lunch	Other	Total
Female	21	13	13	47
Male	19	5	14	38
Total	40	18	27	85

Complete the two-way table.

'School lunch' column: 40 – 21 = 19
'Female' row: 47 – 21 – 13 = 13
'Packed lunch' column: 13 + 5 = 18
'Total' row: 85 – 40 – 18 = 27
'Other' column: 27 – 13 = 14
'Male' row: 19 + 5 + 14 = 38

Check:
47 + 38 = 85
40 + 18 + 27 = 85

Golden rules

The numbers in each column add up to the total for that column.

Other
13
+ 14
= 27

The numbers in each row add up to the total for that row.

Female	21	+ 13	+ 13	= 47

You might have to complete a two-way table in the exam.

1. Look for rows or columns with only one missing number.
2. Use subtraction to find any missing numbers in the table.
3. Use addition to find any missing totals.
4. Fill in the missing values as you go along.

Check it!
Add up the row totals and the column totals. They should be the same.

Now try this

Draw a two-way table for this information. Fill in all the given numbers and then complete the table.

edexcel

1. The two-way table shows some information about the numbers of students in a school.

	Year group			Total
	9	10	11	
Boys			125	407
Girls		123		
Total	303	256		831

Complete the two-way table. **(3 marks)**

2. 80 children went on a school trip. They went to London or to York. 23 boys and 19 girls went to London. 14 boys went to York.

(a) Draw a table or chart to show this information.

(b) How many girls went to York? **(3 marks)**

Stratified sampling

Sample

A SAMPLE is a small group chosen from a larger population.

The red figures represent a sample of 4 from a population of 16.

You can make conclusions about a population by collecting data from a sample.

It is usually cheaper and quicker to collect data from a sample.

Stratified sampling

A stratified sample is one in which the population is split into groups. A simple random sample is taken from each group. The number taken from each group should be in proportion to the size of the group.

There are twice as many boys as girls in this population...

Girls

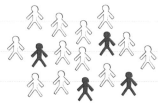
Boys

... so you need twice as many boys as girls in a stratified sample.

Worked example

grade A

The table below gives some information about some students in a school.

Year group	Boys	Girls	Total
Year 12	126	94	220
Year 13	77	85	162
Total	203	179	382

Andrew is going to carry out a survey of these students.

He uses a sample of 50 students, <u>stratified</u> by year group and gender.

Work out the number of Year 13 girls that should be in his sample.

$$\frac{85}{382} \times 50 = 11.12... \approx 11$$

EXAM ALERT!

Only a quarter of students got full marks for this question. If you see the word 'stratified' in a question underline it.

Write the number of Year 13 girls as a fraction of the total population then multiply by the sample size. Round your answer to the nearest whole number.

This was a real exam question that caught students out – **be prepared!**

Now try this

edexcel

1. There are two age groups in a competition. This table shows the number of competitors in each group.

grade A

Age group (years)	Boys	Girls	Total
11–16	105	136	241
17–20	62	90	152
Total	167	226	393

Niki is going to carry out a survey of the competitors.

She uses a sample of 60 competitors, stratified by age and gender.

Work out the number of boys in the 11–16 age group that should be in her sample.   **(2 marks)**

2. The table gives information about the numbers of students in the two years of a college course.

grade A

	Male	Female
First year	399	602
Second year	252	198

Anna wants to interview some of these students.

She takes a random sample of 70 students stratified by year and by gender.

Work out the number of students in the sample who are male and in the first year.   **(3 marks)**

Mean, median and mode

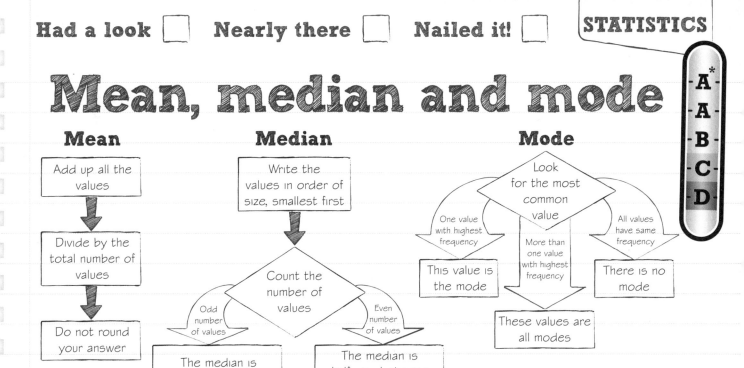

Mean

Add up all the values

↓

Divide by the total number of values

↓

Do not round your answer

Median

Write the values in order of size, smallest first

↓

Count the number of values

Odd number of values → The median is the middle value

Even number of values → The median is half-way between the two middle values

Mode

Look for the most common value

One value with highest frequency → This value is the mode

More than one value with highest frequency → These values are all modes

All values have same frequency → There is no mode

A*·A·A·B·C·D

Worked example

grade **C**

40 people took a driving theory test at a test centre on one day.
10 people took the motorcycle test.
30 people took the car test.
The mean score for all 40 people was 48.2
The mean score for the 30 people who took the car test was 47.7
Calculate the mean score for the 10 people who took the motorcycle test.

Sum of test scores for all 40 people:
$48.2 \times 40 = 1928$

Sum of test scores for 30 people who took car test: $47.7 \times 30 = 1431$

Sum of test scores for 10 people who took motorcycle test: $1928 - 1431 = 497$

Mean score for 10 people who took motorcycle test: $497 \div 10 = 49.7$

To help you solve problems involving the mean, use this formula:

Sum of values = mean × number of values

First you need to calculate the sum of all the test scores for each group. Then:

Sum of test scores for all 40 people
− Sum of test scores for 30 people who took car test
‾‾‾‾‾‾‾‾‾‾‾‾‾‾‾‾‾‾‾‾‾‾‾‾‾‾‾‾‾‾‾‾‾‾‾
Sum of test scores for 10 people who took motorcycle test

Which average works best?

	👍	👎
Mean	Uses all the data	Affected by extreme values
Median	Not affected by extreme values	Value may not exist
Mode	Suitable for data that can be described in words	May not be a mode if data set is small

Now try this

edexcel

1. Jalin wrote down the ages, in years, of seven of his relatives.
45, 38, 43, 43, 39, 40, 39
(a) Write down the mode. **(1 mark)**
(b) Find the median age. **(1 mark)**
(c) Work out the mean age. **(2 marks)**
(d) Which of these three averages best describes this data? Give a reason for your answer. **(1 mark)**

grade **D**

2. A youth club has 60 members.
40 of the members are boys.
20 of the members are girls.
The mean number of videos watched last week by all 60 members was 2.8
The mean number of videos watched last week by the 40 boys was 3.3
Calculate the mean number of videos watched last week by the 20 girls. **(3 marks)**

grade **C**

-A*-
-A-
-B-
-C-
-D-

Frequency table averages

Finding averages from frequency tables is a common exam question.
This frequency table shows the numbers of pets owned by the students in a class.

Number of pets (x)	Frequency (f)	Frequency × number of pets $(f \times x)$
0	12	$12 \times 0 = 0$
1	18	$18 \times 1 = 18$
2	5	$5 \times 2 = 10$
3	2	$2 \times 3 = 6$
Total	37	34

The mode is 1. This value has the highest frequency.

To calculate the mean you need to add a column for '$f \times x$'.

There are 37 values so the median is the $\frac{37 + 1}{2} = $ 19th value.
The first 12 values are all 0. The next 18 values are 1. So the median is 1.

The total in the '$f \times x$' column represents the total number of pets owned by the class.

$$\text{Mean} = \frac{\text{total number of pets}}{\text{total frequency}} = \frac{34}{37} = 0.92 \text{ (to 2 d.p.)}$$

Worked example

grade C

Sethina recorded the times, in minutes, taken to repair 80 car tyres.
Information about these times is shown in the table.

Time (t minutes)	Frequency (f)	Midpoint (x)	$f \times x$
$0 < t \leqslant 6$	15	3	$15 \times 3 = 45$
$6 < t \leqslant 12$	25	9	$25 \times 9 = 225$
$12 < t \leqslant 18$	20	15	$20 \times 15 = 300$
$18 < t \leqslant 24$	12	21	$12 \times 21 = 252$
$24 < t \leqslant 30$	8	27	$8 \times 27 = 216$
	Total frequency = 80		Total of $f \times x$ = 1038

Everything in red is part of the answer.

Calculate an estimate for the mean time taken to repair each car tyre.

$$\text{Mean} = \frac{1038}{80} = 12.975 \approx 13.0 \text{ minutes}$$

Round your answer to 3 significant figures.

EXAM ALERT!

There were 4 marks available for this question but only 40% of students got them all.
Use extra columns in the table for 'Midpoint (x)' and 'Midpoint × frequency ($f \times x$)'.

$$\text{Estimate of mean} = \frac{\text{total of } (f \times x \text{ column})}{\text{total frequency}}$$

This was a real exam question that caught students out – **be prepared!**

 ResultsPlus

Now try this

grade C

The table shows some information about the areas of 50 gardens.

(a) Calculate an estimate for the mean area of these gardens. **(4 marks)**

(b) Explain why the class interval that contains the median is $60 < A \leqslant 80$ **(2 marks)**

Area of garden (A m^2)	Number of gardens (f)
$0 < A \leqslant 20$	4
$20 < A \leqslant 40$	7
$40 < A \leqslant 60$	10
$60 < A \leqslant 80$	22
$80 < A \leqslant 100$	7

Sometimes extra columns in the table are given and sometimes they are not. If not, add them yourself.

Interquartile range

Range and interquartile range are measures of spread. They tell you how spread out data is.

QUARTILES divide a data set into four equal parts.

Half of the values lie between the lower quartile and the upper quartile.

$Q_1 = \dfrac{n+1}{4}$th value, where n = number of data values

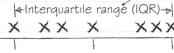

|←Interquartile range (IQR)→|

X X   X   X X   X   XXXX   X   DATA VALUES

Smallest value   Lower quartile (Q₁)   Median (Q₂)   Upper quartile (Q₃)   Largest value

$Q_3 = \dfrac{3(n+1)}{4}$th value

RANGE = largest value − smallest value

INTERQUARTILE RANGE (IQR) = upper quartile (Q₃) − lower quartile (Q₁)

Worked example

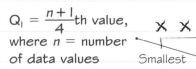

grade **C**

Alison recorded the heights, in cm, of some tree saplings. She put the heights in order.

21   23   23   25   26   26   31   32
33   35   36   40   40   41   42

Work out the interquartile range of Alison's data.

$n = 15$

$\dfrac{n+1}{4} = \dfrac{15+1}{4} = 4$

$Q_1 = $ 4th value $= 25$ cm

$\dfrac{3(n+1)}{4} = \dfrac{3(15+1)}{4} = 12$

$Q_3 = $ 12th value $= 40$ cm

$IQR = Q_3 - Q_1 = 40 - 25 = 15$ cm

To work out the interquartile range, you need to know the lower quartile and the upper quartile.

1. Count the total number of values, n.

2. Check that the data is arranged in order of size.

3. Find the $\dfrac{n+1}{4}$th data value. This is the lower quartile (Q₁).

4. Find the $\dfrac{3(n+1)}{4}$th data value. This is the upper quartile (Q₃).

5. Subtract the lower quartile from the upper quartile to find the interquartile range.

Golden rule

Always arrange the data in order of size before calculating the median or quartiles.
If the data is given in a STEM AND LEAF DIAGRAM then it is already in order of size.
This stem and leaf diagram shows the costs, in £, of some DVDs.

There are 11 pieces of data in this stem and leaf diagram.

This is the STEM.

```
0 | 7  9  9
1 | 0  0  2  3  5  7
2 | 0  5
```

0|9 represents £9

Key: 1 | 5 = £15

There are 11 pieces of data, so the median is the 6th value. The median is £12.

Now try this

edexcel ⠿

Jason collected some information about the heights of 19 plants. This information is shown in the stem and leaf diagram.

(a) Find the median.   **(1 mark)**

(b) Work out the interquartile range of the heights. **(2 marks)**

```
1 | 1  2  3  3
2 | 3  3  5  9  9
3 | 0  2  2  6  6  7
4 | 1  1  4  8
```

Key: 4 | 8 means 48 mm

grade **B**

A* · A · B · C · D

Frequency polygons

You can represent grouped data using a FREQUENCY POLYGON. Look at this example.

Reaction time (r milliseconds)	Frequency
$100 \leqslant r < 200$	7
$200 \leqslant r < 300$	15
$300 \leqslant r < 400$	10

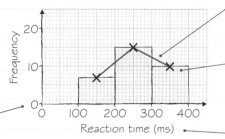

• Join the points with STRAIGHT LINES. Make sure you use a ruler.

• Plot points at the MIDPOINT of each class interval.

• This frequency polygon shows the reaction times of a class of students.

You always record FREQUENCY on the vertical axis.

If you draw a histogram on the same graph the frequency polygon joins together the midpoints of the tops of the bars.

Worked example

grade C

30 students ran a cross-country race.
Each student's time was recorded.
The table gives information about these times.

Time (t minutes)	Frequency	Midpoint
$10 \leqslant t < 14$	2	12
$14 \leqslant t < 18$	5	16
$18 \leqslant t < 22$	12	20
$22 \leqslant t < 26$	8	24
$26 \leqslant t < 30$	3	28

Draw a frequency polygon to show this information.

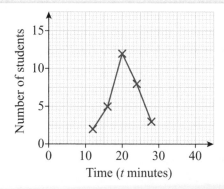

Start by working out the **midpoints** of the class intervals.
The midpoint of the class interval
$10 \leqslant t < 14$ is $\dfrac{10 + 14}{2} = 12$

Check it!
In your exam you will only be asked to draw a frequency polygon for data with **equal class intervals**. So make sure that your midpoints are the same distance apart.

Now try this

grade C

edexcel

60 students take a geography test.
The test is marked out of 50

This table gives information about the students' marks.

Geography mark	0−10	11−20	21−30	31−40	41−50
Frequency	5	11	19	16	9

On the grid, draw a frequency polygon to show this information. **(2 marks)**

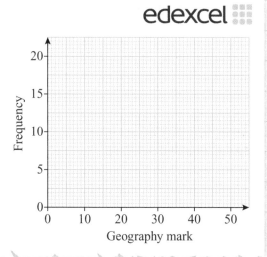

Histograms

-A*-
-A-
-B-
-C-
-D-

Histograms are usually used to represent grouped data with different class widths.

Worked example

 grade A

The incomplete table and histogram give information about the ages of the people who live in a village.

Age (x) in years	Frequency
$0 < x \leqslant 10$	160
$10 < x \leqslant 25$	$4 \times 15 = 60$
$25 < x \leqslant 30$	$8 \times 5 = 40$
$30 < x \leqslant 40$	100
$40 < x \leqslant 70$	120

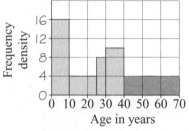

Everything in red is part of the answer.

(a) Use the information in the histogram to complete the frequency table.

Frequency density of $0 < x \leqslant 10$ class

$= \dfrac{160}{10} = 16$

(b) Complete the histogram.

Frequency density of $40 < x \leqslant 70$ class

$= \dfrac{120}{30} = 4$

Histogram facts

No gaps between the bars. ✓

Area of each bar is proportional to frequency. ✓

Vertical axis is labelled 'Frequency density'. ✓

Bars can be different widths. ✓

Frequency density $= \dfrac{\text{frequency}}{\text{class width}}$ ✓

(a) You know the frequency for the $0 < x \leqslant 10$ class.
Use this to work out the scale on the vertical axis.

Use 'frequency = frequency density × class width' to work out the missing frequencies.

(b) Alternatively, you could also 'count squares' to complete the histogram. Work out how many people are represented by one square.

You can use the area under a histogram to estimate frequencies.

An estimate for the number of maggots between 1 mm and 2 mm long is:

$0.5 \times 22 + 0.5 \times 6 = 11 + 3 = 14$

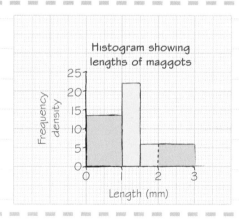

Histogram showing lengths of maggots

Now try this

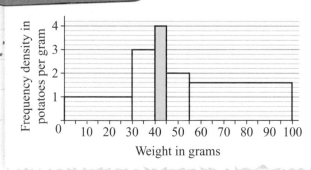

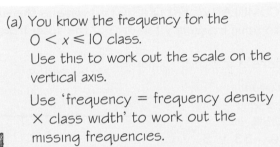

edexcel

The histogram gives information about the weights of some potatoes.
The shaded bar represents 20 potatoes.

grade A

(a) Work out how many of the potatoes weigh 30 grams or less. **(1 mark)**

grade A*

(b) Estimate the number of potatoes which weigh more than 50 grams. **(2 marks)**

Cumulative frequency

In your exam you might have to draw a cumulative frequency graph, or use one to find the median or the interquartile range.

How to draw a cumulative frequency graph

Reaction time t (s)	Frequency	Cumulative frequency
$0 < t \leqslant 0.1$	2	2
$0.1 < t \leqslant 0.2$	5	$2 + 5 = 7$
$0.2 < t \leqslant 0.3$	18	$7 + 18 = 25$
$0.3 < t \leqslant 0.4$	5	$25 + 5 = 30$
$0.4 < t \leqslant 0.5$	1	$30 + 1 = .31$

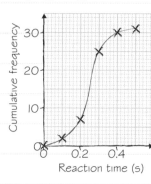

1. Plot 0 at the beginning of the first class interval.

2. Plot each value at the UPPER end of its class interval.

3. Join your points with a SMOOTH CURVE.

Add a column for CUMULATIVE FREQUENCY to your frequency table.

Check that your final value is the same as the total frequency.

Here's another example:

Cumulative frequency diagram of test results

33 students scored less than 75%.
So $36 - 33 = 3$ students scored more than 75%.

The interquartile range is
$64\% - 42\%$
$= 22\%$

Draw the lower quartile at cumulative frequency $= \dfrac{36}{4}$
The lower quartile was 42%.

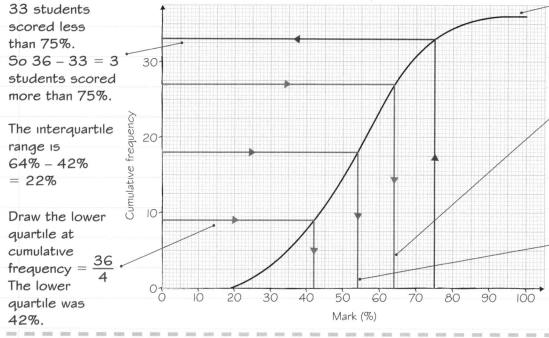

There were 36 students in the class. (This is the FIRST FACT you should establish.)

Draw the upper quartile at cumulative frequency $= \dfrac{3 \times 36}{4}$
The upper quartile was 64%.

Draw the median at cumulative frequency $= \dfrac{36}{2}$
The median was 54%.

Now try this

grade **B**

edexcel

An operator took 100 calls at a call centre. The table gives information about the time it took the operator to answer each call.

(a) Complete the cumulative frequency column. **(1 mark)**

(b) Draw a cumulative frequency graph for your table. **(2 marks)**

(c) Use your graph to find an estimate for the interquartile range of the times. **(2 marks)**

(d) Use your graph to find an estimate for the number of calls the operator took more than 18 seconds to answer. **(2 marks)**

Time (t seconds)	Frequency	Cumulative frequency
$0 < t \leqslant 10$	16	16
$10 < t \leqslant 20$	34	
$20 < t \leqslant 30$	32	
$30 < t \leqslant 40$	14	
$40 < t \leqslant 50$	4	

Box plots

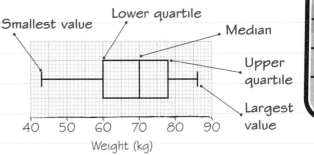

A*
A
B
C
D

Box plots show the median, upper and lower quartiles, and the largest and smallest values of a set of data. They are often used to compare distributions.

Half the weights were between 60 kg and 78 kg. 25% of the weights were greater than 78 kg.

Smallest value   Lower quartile   Median   Upper quartile   Largest value
Weight (kg)
40 50 60 70 80 90

Worked example
grade B

The box plot gives information about the heights, in metres, of some trees.

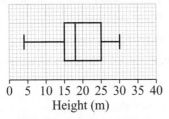

0 5 10 15 20 25 30 35 40
Height (m)

Fred says that exactly 50% of the trees have heights between 15 m and 30 m.
Is Fred correct? You must give a reason.

15 m → lower quartile (25%)
30 m → largest value (100%)
No, Fred is not correct, because 75% of the trees have heights between 15 m and 30 m.

EXAM ALERT!

There was 1 mark available here but only one in four students got it. You have to give a reason for your answer. 'Yes' or 'No' would score no marks.

Use the box plot to identify what the heights 15 m and 30 m represent. Then use these facts to decide whether Fred is correct.

This was a real exam question that caught students out – **be prepared!**
Results Plus

Comparing two distributions

When you are comparing box plots, you can use the range, the interquartile range, the largest and smallest values and the median.

It is always best to compare measures of spread rather than medians or end points.

Now try this
grade B

The cumulative frequency graph below gives information about the times taken by 40 boys to complete a puzzle.

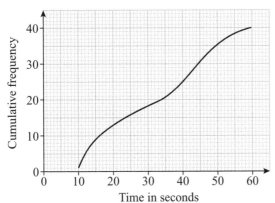

Cumulative frequency
40
30
20
10
0
0 10 20 30 40 50 60
Time in seconds

For the boys the minimum time to complete the puzzle was 9 seconds and the maximum time to complete the puzzle was 57 seconds.

Read off the largest and smallest values, the median and the quartiles from the graph.

edexcel

(a) Use this information and the cumulative frequency graph to draw a box plot showing information about the boys' times. **(3 marks)**

The box plot below shows information about the times taken by 40 girls to complete the same puzzle.

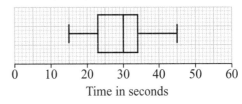

0 10 20 30 40 50 60
Time in seconds

(b) Make two comparisons between the boys' times and the girls' times. **(2 marks)**

A*
A
B
C
D

Scatter graphs

The points on a scatter graph aren't always scattered. If the points are almost on a straight line then the scatter graph shows CORRELATION. The better the straight line, the stronger the correlation.

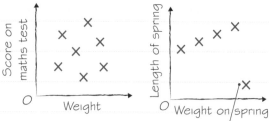

Negative correlation

No correlation

Positive correlation

An isolated point on a scatter graph is an extreme point that lies outside the normal range of values.

Worked example

grade D

A garage sells motorcycles.

The scatter graph gives information about the ages and prices of the motorcycles.

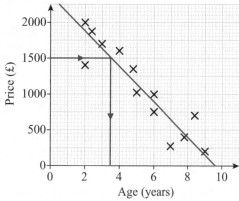

(a) What type of correlation does the scatter graph show?

Negative

(b) Draw a line of best fit on the scatter graph.

Mae buys a motorcycle from the garage for £1500

(c) Use your line of best fit to estimate the age of the motorcycle.

3.5 years

Line of best fit checklist

Straight line that is as close as possible to all the points. ✓

Used to predict values. ✓

Does not need to go through (0, 0). ✓

Drawn with a ruler. ✓

Ignores isolated points. ✓

Remember that the type of correlation tells you about the relationship between price and age. Negative correlation means that as the age increases the price decreases.

To predict the age of the motorcycle, read across from £1500 on the vertical axis then down to the horizontal axis. Draw the lines you use on your graph.

Now try this

edexcel

grade D

The scatter graph gives information about the area and the cost of some pictures.

All the pictures are rectangles.

The line of best fit has been drawn on the graph.

One of the pictures costs £1000. Its length is 48 cm.

Use the line of best fit to estimate the width of the picture. **(3 marks)**

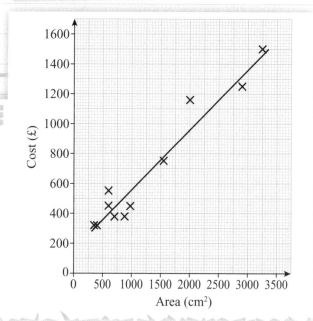

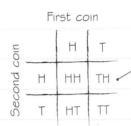

-A*-
-A-
-B-
-C-
-D-

Probability

Here are four things you need to know about basic probability for your exam.

1 A probability is the chance an event will happen, and is always a number from 0 to 1.

For EQUALLY LIKELY OUTCOMES, the probability (P) that something will happen is:

$$\text{Probability} = \frac{\text{number of successful outcomes}}{\text{total number of outcomes}}$$

If you know the probability that something will happen, you can calculate the probability that it won't happen.

P(event doesn't happen)
= 1 − P(event does happen)

2 A sample space diagram shows you all the possible outcomes of an event.

Here are all the possible outcomes when two coins are flipped.

First coin

	H	T
H	HH	TH
T	HT	TT

Second coin

There are four possible outcomes. TH means getting a Tail on the first coin and a Head on the second coin.

3 RELATIVE FREQUENCY is sometimes called EXPERIMENTAL PROBABILITY.

You can estimate a probability using the results of an experiment.

$$\text{Estimated probability} = \frac{\text{number of successful trials}}{\text{total number of trials}}$$

If you calculate a probability using relative frequency, it is only an ESTIMATE.

The more trials you carry out, the more accurate your estimate will be.

4 Probability helps you predict the outcome of an event.

If you flip a coin 100 times, you can EXPECT to throw Heads about 50 times.

You might not throw Heads exactly 50 times, but it's a good guess.

Expected number of outcomes
= number of trials × probability

Worked example

Amir designs a game for his school fete.

It costs 80p to play.

The probability of winning the game is $\frac{2}{5}$

The prize for winning is £1.50

200 people play Amir's game.

Work out an estimate of the profit Amir should expect to make.

grade **C**

80p = £0.80
Money taken in total
= 200 × £0.80 = £160
Expected number of winners
= 200 × $\frac{2}{5}$ = 80
Money paid in prizes
= 80 × £1.50 = £120
Expected profit = £160 − £120 = £40

Now try this

edexcel

grade **D**

1. The probability that a biased dice will land on a 4 is 0.2
Pam is going to roll the dice 300 times.
Work out an estimate for the number of times the dice will **not** land on a 4
(3 marks)

2. Jon designs a game to raise money for charity.
It costs £2 to play.
The probability of winning the game is $\frac{1}{5}$
The prize for winning is £5
400 people play Jon's game.
Work out an estimate of the profit Jon should expect to make.
(3 marks)

grade **C**

A* A B C D

Tree diagrams

A tree diagram shows all the possible outcomes from a series of events and their probabilities.

This is a tree diagram for Holly's journey to school.

You write the probability for each event on the branch.

At each branch the probabilities add up to 1.
$\frac{2}{3} + \frac{1}{3} = 1$

The outcome of the first event can affect the probability of the second.
Holly is less likely to be on time if she misses the bus.

Catch bus $\frac{2}{3}$ — $\frac{1}{4}$ Late

$\frac{3}{4}$ On time

Miss bus $\frac{1}{3}$ — $\frac{4}{5}$ Late

$\frac{1}{5}$ On time

Each branch is like a different parallel universe. In this universe, Holly catches the bus and gets to school on time.

You write the outcomes at the ends of the branches.
You can use shorthand like this.

Outcome	Probability
CL	$\frac{2}{3} \times \frac{1}{4} = \frac{2}{12} = \frac{1}{6}$
CO	$\frac{2}{3} \times \frac{3}{4} = \frac{6}{12} = \frac{1}{2}$
ML	$\frac{1}{3} \times \frac{4}{5} = \frac{4}{15}$
MO	$\frac{1}{3} \times \frac{1}{5} = \frac{1}{15}$

You multiply along the branches to find the probability of each outcome.
The probability that Holly misses the bus and is late for school is $\frac{4}{15}$.

Golden rules

1 Look out for the words REPLACE or PUT BACK in a probability question.

WITH replacement: probabilities stay the same.

WITHOUT replacement: first probability stays the same while the others change.

2 MULTIPLY ALONG THE BRANCHES × ✚ ADD UP THE OUTCOMES

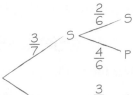

Worked example

grade A*

There are 3 strawberry yoghurts and 4 peach yoghurts in a fridge. Kate takes a yoghurt at random from the fridge. She eats the yoghurt. She then takes a second yoghurt at random from the fridge. Work out the probability that both the yoghurts were the same flavour.

First yoghurt	Second yoghurt	Outcome	Probability
$\frac{3}{7}$ S	$\frac{2}{6}$ S	SS	$\frac{3}{7} \times \frac{2}{6} = \frac{1}{7}$
	$\frac{4}{6}$ P	SP	$\frac{3}{7} \times \frac{4}{6} = \frac{2}{7}$
$\frac{4}{7}$ P	$\frac{3}{6}$ S	PS	$\frac{4}{7} \times \frac{3}{6} = \frac{2}{7}$
	$\frac{3}{6}$ P	PP	$\frac{4}{7} \times \frac{3}{6} = \frac{2}{7}$

P (both yoghurts same flavour) = P(SS) + P(PP)
$= \frac{1}{7} + \frac{2}{7} = \frac{3}{7}$

EXAM ALERT!

Use a tree diagram to answer question 1 below. Only one in ten students got full marks for this question in the exam.

This was a real exam question that caught students out – **be prepared!**

 ResultsPlus

Now try this

 edexcel

1. Fred has a biased coin. The probability of getting Heads on one throw of the coin is $\frac{3}{4}$. He throws the biased coin 3 times. Work out the probability that he gets at least two Heads. **(3 marks)**

grade A*

2. There are 4 bottles of orange juice, 3 bottles of apple juice, and 2 bottles of tomato juice. Viv takes a bottle at random and drinks the juice. Then Caroline takes a bottle at random and drinks the juice. Work out the probability that they both take a bottle of the same type of juice. **(4 marks)**

grade A*

Problem-solving practice

About half of the questions on your exam will need problem-solving skills.

These skills are sometimes called AO2 and AO3.

Practise using the questions on the next two pages.

For these questions you might need to:

- choose which mathematical technique or skill to use
- apply a technique in a new context
- plan your strategy to solve a longer problem
- show your working clearly and give reasons for your answers.

---

1 A teacher asked 30 students if they had a school lunch or a packed lunch or if they went home for lunch.

17 of the students were boys.

4 of the boys had a packed lunch.

7 girls had a school lunch.

3 of the 5 students who went home were boys.

Work out the number of students who had a packed lunch. **(4 marks)**

Two-way tables p. 91   grade **D**

You could get in a real mess with this question unless you draw a two-way table like this. Fill in all the given values and then complete the table.

	School lunch	Packed lunch	Home for lunch	Total
Boys				
Girls				
Total				

TOP TIP

After you've filled in your table, double-check that all the information in the question matches what you have in your table.

---

2 *Some students in a class weighed themselves.

Here are their results.

Boys' weights in kg
70 65 45 52 63 72 63

Girls' weights in kg
65 45 47 61 44 67 55 56 63

Compare fully the weights of these students. **(6 marks)**

Mean, median and mode p. 93   grade **D**

To give a full answer, you need to **compare** the data. So
(1) calculate averages like the mean or median and a measure of spread like the range
(2) write a sentence for each of these, comparing the boys and the girls.

TOP TIP

If a question has a * next to it, then there are marks available for QUALITY OF WRITTEN COMMUNICATION. This means you must show all your working and write your answer clearly with the correct units.

103

Problem-solving practice

3 The box plot gives information about the distribution of the weights of bags on a plane.

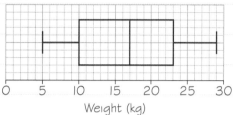

Weight (kg)

Jean says the heaviest bag weighs 23 kg.

Is Jean correct? You must give a reason.   (1 mark)

Box plots p. 99

grade **B**

Writing 'yes' or 'no' is not a full answer. You have to give a reason **using the information in the box plot.**

TOP TIP

When you are giving reasons, you should write down values from a graph or table as evidence.

4 The table and histogram show some information about the areas of some carpets.

Carpet area (A m²)	Frequency
10 < A ≤ 12	
12 < A ≤ 15	15
15 < A ≤ 20	12
20 < A ≤ 25	10

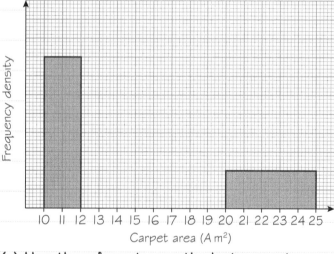

Carpet area (A m²)

Histograms p. 97

grade **A**

Your first job is to work out the scale on the vertical axis. You are given the bar for the class interval 20 < A ≤ 25. Work out the frequency density for this class and write this value as the height of the last bar. You can now fill in the rest of the scale on the vertical axis and calculate the frequency for the 10 < A ≤ 12 class interval.

TOP TIP

You will always need to use this rule in a question on histograms:

$$\text{frequency} = \frac{\text{frequency}}{\text{density}} \times \frac{\text{class}}{\text{width}}$$

(a) Use the information in the histogram to complete the table.   (2 marks)

(b) Use the information in the table to complete the histogram.   (2 marks)

5 Phil has 20 sweets in a bag.

5 of the sweets are orange.

7 of the sweets are red.

8 of the sweets are yellow.

Phil takes TWO sweets at random from the bag.

Work out the probability that the sweets will NOT be the same colour.   (4 marks)

Tree diagrams p. 102

grade **A***

If you draw a tree diagram, you will probably get some or all of the marks for this question because you won't leave out any of the possibilities.

TOP TIP

The sweets are **not replaced** so the probabilities change for the second sweet.

Formulae page

Volume of a prism = area of cross section × length

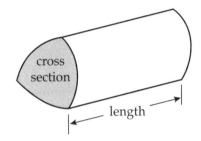

Area of trapezium = $\frac{1}{2}(a + b)h$

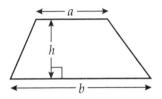

Volume of sphere = $\frac{4}{3}\pi r^3$

Surface area of sphere = $4\pi r^2$

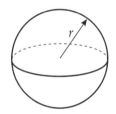

Volume of cone = $\frac{1}{3}\pi r^2 h$

Curved surface area of cone = $\pi r l$

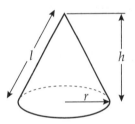

In any triangle *ABC*

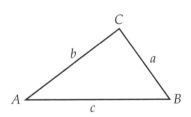

Sine Rule $\dfrac{a}{\sin A} = \dfrac{b}{\sin B} = \dfrac{c}{\sin C}$

Cosine Rule $a^2 = b^2 + c^2 - 2bc \cos A$

Area of triangle = $\frac{1}{2}ab \sin C$

The Quadratic Equation

The solutions of $ax^2 + bx + c = 0$ where $a \neq 0$, are given by

$$x = \frac{-b \pm \sqrt{(b^2 - 4ac)}}{2a}$$

Answers

The number given to each topic refers to its page number.

NUMBER

1. Factors and primes
(a) (i) $2 \times 2 \times 3 \times 5$ (ii) $2 \times 2 \times 2 \times 2 \times 2 \times 3$
(b) 12   (c) 480

2. Indices 1
1. (a) 81   (b) 8
2. (a) a^8 (b) 3

3. Fractions
(a) $4\frac{5}{12}$ inches   (b) $1\frac{31}{32}$

4. Decimals
1. (a) 14.95   (b) 149.5
2. (a) 119.31   (b) 119 310   (c) 1.23

5. Recurring decimals
1. $n = 0.181818...$; $100n = 18.181818...$; $99n = 18$; $n = \frac{18}{99}$
2. (a) $\frac{36}{99}$ or equivalent   (b) $2\frac{3}{22}$

6. Rounding and estimation
1. 12 000
2. 20

7. Upper and lower bounds
8.75 miles per litre

8. Fractions and percentages
(a) 62.5%
(b) $\frac{1}{4}$

9. Percentage change
1. £60.16
2. 5%

10. Reverse percentages and compound interest
1. £4867.20
2. $n = 5$

11. Ratio
1. £60
2. sugar 160 g, butter 240 g, flour 400 g

12. Proportion
1. £32.89
2. 300 g

13. Indices 2
1. (a) $\frac{1}{4}$ (b) -2
2. (a) (i) $\frac{1}{9}$ (ii) ± 6 (iii) 9   (iv) $\frac{27}{8}$ or equivalent
   (b) $\frac{1}{4}$

14. Standard form
(a) 8.06×10^4 (b) 6×10^9

15. Calculator skills
1. 0.928 739 729 …
2. 2.31×10^{10}

16. Surds
(a) $3\sqrt{2}$ (b) $9\,\text{cm}^2$

17–18. Problem-solving practice
1. 18 days
2. Area of field = 23 750 m²
Farmer Boyce will offer £23 750.
Farmer Giles's offer is better.
3. 24.72 litres
4. $n = 4$
5. 75%

ALGEBRA

19. Algebraic expressions
1. (a) p^9 (b) x^5 (c) q^6 (d) $12s^6t^5$
2. (a) (i) k^3 (ii) m^8 (iii) $6r^3t^6$ (iv) $81x^4y^8$
   (b) $A = 4, m = 6, n = 6$

20. Arithmetic sequences
(a) $4n - 1$
(b) 318 is even and all the terms in the sequence are odd
(c) 3, 1, -1

21. Expanding brackets
1. (a) $35 - 14x$ (b) $16x + 42$ (c) $y^2 + y - 12$
2. $2x^2 - x - 15$
3. (a) $2x^2 - 7xy - 15y^2$ (b) $26x - 13$

22. Factorising
(a) $6x(2x - 3y)$ (b) $(a + 3b)(a - 3b)$ (c) $3x(2x - 3y)$
(d) $(2x - 3)(x - 2)$ (e) $(p + q)(p + q + 5)$ (f) $2(3f - 2)(f - 1)$

23. Linear equations 1
1. (a) $\frac{1}{2}$ (b) $\frac{2}{5}$
2. $\frac{7}{2}$ or $3\frac{1}{2}$

24. Linear equations 2
1. $\frac{3}{8}$
2. -3

25. Straight-line graphs
$y = -4x + 17$

26. Parallel and perpendicular
(a) $y = \frac{1}{2}x + k$
(b) $y = mx + 1, m \neq \frac{1}{2}$
(c) $y = -2x + 26$

27. 3-D coordinates
(a) B (b) $(1, \frac{1}{2}, 1\frac{1}{2})$

28. Real-life graphs
(a) 30 km   (b) 45 minutes   (c) 60 km/h

29. Formulae
(a) -5 (b) $N = 5r + 3g$

30. Rearranging formulae
1. $s = \dfrac{v^2 - u^2}{2a}$
2. $x = \dfrac{4y + 15}{5 + 3y}$

31. Inequalities
1. $-1, 0, 1$ 2. (a) $x > \frac{1}{7}$ (b) 1

32. Inequalities on graphs

Graph with crosses at $(-2, 1), (-2, 2), (-2, 3), (-1, 1), (-1, 2), (0, 1)$

33. Quadratic and cubic graphs

(a) $6, -2, 0$

(b)

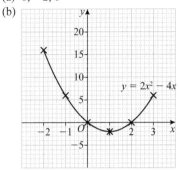

(c) 2.6 or -0.6

34. Graphs of $\frac{k}{x}$ and a^x

$k = \frac{1}{4}$

35. Trial and improvement

1. 2.7

2. 3.6

36. Simultaneous equations 1

1. $x = -1, y = 1.5$

2. $x = 6, y = -5$

37. Quadratic equations

(a) $\left(\dfrac{x + 2 + x + 6}{2}\right), (x + 4)(x - 5), x^2 - 5x + 4x - 20,$
$x^2 - x - 20 = 36$

(b) (i) $x = 8$ or -7 (ii) $3\,\text{cm}$

38. Completing the square

1. $x^2 - 4x + 15 = (x - 2)^2 - 4 + 15 = (x - 2)^2 + 11$
    $p = -2, q = 11$

2. $-2 \pm \sqrt{\dfrac{11}{2}}$

39. The quadratic formula

1. $x = 1.22$ or -3.55

2. $x = 1.30$ or -2.30

40. Quadratics and fractions

1. $x = 1.27$ or -6.27

2. $x = -3.5$

41. Equation of a circle

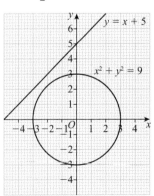

As the graphs will never meet there are no points that satisfy both equations.

Hence there are no solutions to the simultaneous equations.

42. Simultaneous equations 2

1. $x = 2, y = 5$ or $x = -5, y = -2$

2. $x = 2, y = 3$ or $x = -1, y = 0$

43. Direct proportion

(a) $d = 5t^2$ (b) $45\,\text{m}$ (c) $11\,\text{s}$

44. Proportionality formulae

$q = 1.2$

45. Transformations 1

(a) (i) $(0, -1)$ (ii) $(2, -3)$ (iii) $(1, -1)$

(b) $y = \text{f}(-x)$

46. Transformations 2

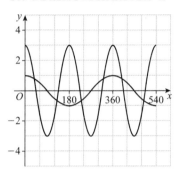

47. Algebraic fractions

1. $\dfrac{7x - 11}{12}$

2. (a) $\dfrac{x}{x - 5}$ (b) $\dfrac{x + 1}{2x + 3}$

48. Proof

e.g. $(2n + 3)^2 - (2n + 1)^2 = 4n^2 + 12n + 9 - (4n^2 + 4n + 1)$
$= 8n + 8 = 8(n + 1)$, which is divisible by 4

49–50. Problem-solving practice

1. $t = 4\frac{1}{3}$

2. B has coordinates $(-7, -10, 8)$

3. $k = 1\frac{1}{6}$

4. $V = 10x^2 + 24x - 18$

5. (a)

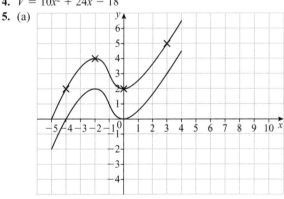

(b) A reflection in the x-axis

GEOMETRY AND MEASURES

51. Angle properties

Let angle $BAC = x$

angle BAC = angle ACD	(alternate angles equal)
angle BAC = angle ACB	(base angles of an isosceles triangle are equal)
angle ACD = angle ACB	(both equal to angle BAC)
so AC bisects angle DCB	

52. Solving angle problems

(a) (i) 62°   (ii) alternate angles are equal
(b) angle BEF = angle BFE = 62°
    (base angles of an isosceles triangle are equal)
    angle EBF = 180 − 62 − 62 = 56°
    (angles in a triangle add up to 180°)

53. Angles in polygons

(a) 9
(b) The sum of the exterior angles of a polygon is 360°.
    Each exterior angle of the hexagon is 360 ÷ 6 = 60°.
    angle AGF = 180 − 60 − 60 = 60°
    (angles in a triangle add up to 180°)

54. Plan and elevation

(a)   (b)

55. Perimeter and area

20 (area = 102 m²)

56. Prisms

(a) $\frac{1}{2}(4 \times 3) + \frac{1}{2}(4 \times 3) + (7 \times 5) + (7 \times 3) + (7 \times 4)$
   = 6 + 6 + 35 + 21 + 28 = 96 cm²
(b) 42 cm³

57. Circles and cylinders

(a) 7854 cm²   (b) 126 cm

58. Sectors of circles

37.7 cm²

59. Volumes of 3-D shapes

905 m³

60. Pythagoras' theorem

49.4 cm²

61. Surface area

(a) 5 cm   (b) 78.5 cm²

62. Converting units

Jane: 1 hour 36 minutes   Mattie: 1 hour 40 minutes

63. Units of area and volume

1. (a) 25 000 cm²   (b) 8.56 cm²
   (c) 7000 mm³   (d) 0.856 m³
2. 30 litres

64. Speed

(a) 09:30   (b) 37.5 mph

65. Density

302 g

66. Congruent triangles

$PQ = PR$ (given)
$PY = PX$ (given)
angle P = angle P (common)
SAS

67. Similar shapes 1

(a) 8 cm   (b) 12.5 cm

68. Similar shapes 2

1. 3200 cm³
2. 3800 cm²

69. Bearings

(a)

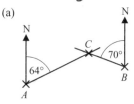

(b) 308°

70. Scale drawings and maps

(a) 076°
(b)

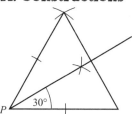

(c) 29.2 km

71. Constructions

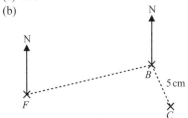

72. Loci

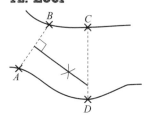

73. Translations, reflections and rotations

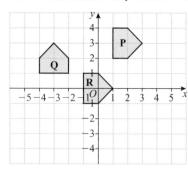

74. Enlargements

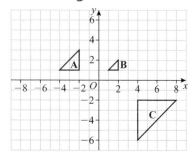

75. Combining transformations

Rotation by 180° about (2, −1)

76. Line segments

$4\sqrt{5}$

77. Trigonometry 1

33.9°

78. Trigonometry 2

1. 5.66 cm
2. 12.7 cm

79. Pythagoras in 3-D

9.11 cm

80. Trigonometry in 3-D

(a) 11.2 m   (b) 15.6°

81. Triangles and segments

9.51 cm²

82. The sine rule

41.9°

83. The cosine rule

5.89 cm

84. Circle facts

angle $LMO = 90°$ (angle between a radius and a tangent is 90°)
angle $MON = (180 − 2x)°$ (angles on a straight line add up to 180°)
angle $NMO = [180 − (180 − 2x)] ÷ 2$
$= 2x ÷ 2$
$= x°$ (isosceles triangle, so base angles are equal)
angle $LMN = (90 − x)°$

85. Circle theorems

(a) (i) 140   (ii) angle at centre is twice angle at circumference
(b) (i) 110   (ii) opposites angles in a cyclic quadrilateral sum to 180°

86. Vectors

(a) $2\mathbf{a} + 2\mathbf{b}$ (b) $7\mathbf{a} + 6\mathbf{b}$

87. Solving vector problems

(a) (i) $\frac{1}{2}\mathbf{a}$ (ii) $\frac{1}{2}\mathbf{a} − \frac{1}{2}\mathbf{c}$
(b) $\overrightarrow{CA} = \mathbf{a} − \mathbf{c}$ $\overrightarrow{MN} = \frac{1}{2}(\mathbf{a} − \mathbf{c})$ $\overrightarrow{MN} = \frac{1}{2}\overrightarrow{CA}$
Since $\overrightarrow{MN}$ is a multiple of $\overrightarrow{CA}$, the lines MN and CA must be parallel.

88–89. Problem-solving practice

1. For example:
   $\angle ABE = 65°$ (opposite angles in parallelogram are equal)
   $x = 115°$ (angles on a straight line add up to 180°)
2. 8 flowerpots
3. 5.8 m
4. 8580 kg/m³ (to 3 s.f.)
5. 31.7 cm²

STATISTICS AND PROBABILITY

90. Collecting data

(a) e.g. overlapping regions, no time frame
(b) How many hours, to the nearest hour, did you use your mobile phone last week?

   ☐ 0–1   ☐ 2–3   ☐ 4–5   ☐ more than 5

91. Two-way tables

1.

	Year group			Total
	9	10	11	
Boys	149	133	125	407
Girls	154	123	147	424
Total	303	256	272	831

2. (a)

	London	York	Total
Boys	23	14	(37)
Girls	19	(24)	(43)
Total	(42)	(38)	80

(b) 24

92. Stratified sampling

1. 16
2. 19

93. Mean, median and mode

1. (a) 39 and 43   (b) 40   (c) 41
        (d) mean as ages quite close to each other
2. 1.8

94. Frequency table averages

(a) 58.4 m²   (b) 25th value falls in class interval $60 < A \leqslant 80$

95. Interquartile range

(a) 30 mm   (b) 14 mm

96. Frequency polygons

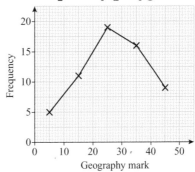

97. Histograms

(a) 30   (b) 82

98. Cumulative frequency

(a) (16), 50, 82, 96, 100   (b) accurately drawn graph
(c) approx. 15 seconds   (d) about 55

99. Box plots

(a)
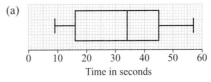

(b) Boys' IQR (29 s) > Girls' IQR (11 s)
     Boys' median (34 s) > Girls' median (30 s)

100. Scatter graphs

43.7 cm–44.8 cm

101. Probability

1. 240
2. £400

102. Tree diagrams

1. $\frac{27}{32}$ or equivalent
2. $\frac{20}{72}$ or equivalent

103–104. Problem-solving practice

1. 8
2. For example:

	Mean	Median	Range
Boys	61.4 kg	63 kg	27 kg
Girls	55.9 kg	56 kg	23 kg

The boys had a higher mean and median than the girls, so they were on average heavier.

The boys had a higher range than the girls, so their weights were more spread out.

3. For example:

No, the heaviest bag weighs 30 kg.

4. (a)

Carpet area (A m²)	Frequency
$10 < A \leqslant 12$	16
$12 < A \leqslant 15$	15
$15 < A \leqslant 20$	12
$20 < A \leqslant 25$	10

(b)

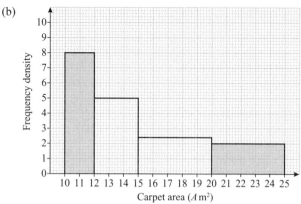

5. $\frac{131}{190}$

Published by Pearson Education Limited, a company incorporated in England and Wales, having its registered office at Edinburgh Gate, Harlow, Essex, CM20 2JE. Registered company number: 872828

www.pearsonschoolsandfecolleges.co.uk

Text © Keith Pledger, Harry Smith and Pearson Education Limited 2011
Edited by Fiona McDonald and Laurice Suess
Typeset by Tech-Set Ltd, Gateshead
Original illustrations © Pearson Education Limited 2011

The rights of Keith Pledger and Harry Smith to be identified as authors of this work have been asserted by them in accordance with the Copyright, Designs and Patents Act 1988.

First published 2011

18 17 16 15
22 21 20

British Library Cataloguing in Publication Data
A catalogue record for this book is available from the British Library

ISBN 978 1 44690 018 5

Printed in Slovakia by Neografia

Disclaimer
This material has been published on behalf of Edexcel and offers high-quality support for the delivery of Edexcel qualifications.
This does not mean that the material is essential to achieve any Edexcel qualification, nor does it mean that it is the only suitable material available to support any Edexcel qualification. Material from this publication will not be used verbatim in any examination or assessment set by Edexcel. Any resource lists produced by Edexcel shall include this and other appropriate resources.

Copies of official specifications for all Edexcel qualifications may be found on the Edexcel website: www.edexcel.com